From

Russell C. Derben

BIOLOGICAL RECLAMATION of Solid Wastes

BIOLOGICAL RECLAMATION of Solid Wastes

Clarence G.Golueke

 Rodale Press, Emmaus, PA

Printed in the United States of America on recycled paper.

Library of Congress Cataloging in Publication Data

Golueke, C.G. 1917-
    Biological reclamation of solid wastes.

    Includes index.
    1. Compost.   2. Refuse and refuse disposal.
3. Sewage-Purification—Biological treatment.
I. Title.
TD796.5.G64   628'.445   77-5564
ISBN 0-87857-158-2

2   4   6   8   10   9   7   5   3   1

Contents

2. LAND DISPOSAL OF PRIMARY AND SECONDARY DOMESTIC SEWAGE SLUDGES

List of Figures

List of Tables

INTRODUCTION

Open dumping, burial, submersion in the ocean, incineration, and combinations thereof continue to be the conventional procedures for municipal solid waste disposal. Happily in the United States the open dump and disposal in the ocean are gradually being phased out. On the other hand, burial will always be a necessity to some extent, because every other method of treatment leaves a residue that can be disposed of only by burial in the land (landfilling). These methods have two things in common: no resources are recovered, and they are physical in nature. However, in the offing is a collection of physical, chemical, and biological methods of solid waste treatment that are designed for recovery of resources.

The only biological method of solid waste treatment that has been developed beyond the pilot plant stage thus far is composting. But, until quite recently an unfortunate combination of circumstances has resulted in the failure of composting to be considered in the United States as a viable method of municipal solid waste treatment other than for park and garden debris and some agricultural wastes. The situation has been only slightly better in countries other than the United States. Aside from composting, biological systems with a potential for processing wastes on a large scale are the spreading of wastewater sludges on the land, anaerobic digestion, and hydrolysis of cellulosic wastes to sugars and subsequent production of single-cell protein or ethanol. A key feature common to these four systems is the recovery of one or more resources. However, of the four, only land spreading and composting hitherto have been applied on a scale not exceeding the small pilot plant.

Fortunately, the situation as far as composting is concerned is beginning to change because the factors responsible for the recent trend back to biological systems in the treatment of wastewaters are beginning to be recognized as applicable to the treatment of solid wastes. Among the factors is the very important one of a generally lesser energy requirement for biological systems than for those that are chemical or physical in nature. Reduced energy needs, coupled with less expensive equipment requirements, make biological systems economically more attractive in these days of energy shortages and inflation of the economy. Another important factor is the fact that their impact on the environment is far less unfavorable than even the better controlled physical or chemical processes, inasmuch as biological treatment processes are but carefully controlled and accelerated applications of those that occur in nature. Finally, it may be said that generally a more complete conservation of resources is possible with biological approaches. With respect to solid waste management, however, two qualifications of these reasons (i.e., the advantages of biological systems) must be kept in mind: (1) Biological processes are applicable only to the organic fraction of solid wastes. (2) Even with all other resources recovered, a residue is left which can be satisfactorily disposed of only by burial. In municipal refuse management, the necessary restriction of a biological system to the organic fraction can be made to serve as a strong incentive for conserving inorganic resources. The rationale is that since the inorganics must be removed, a second step to sort out useful components from the inorganic fraction could easily be made.

Although the rising interest in biological processes for solid waste treatment is leading to the generation of an abundance of literature on the subject, the publications are scattered in a wide array of periodicals, reports, and proceedings of conferences. The problem is compounded by the fact that the majority of the reports are difficult to obtain. Nor is the problem alleviated by the policy of the federal government of charging substantial prices for publications (e.g., NTIS) issued by it. On the other hand, the average textbook on solid waste management stresses the conventional technology, and rightly so in view of its space limitations, and gives scant attention to biological systems.

Consequently, it was felt that a description and discussion of the principal biological processes for solid wastes systems for solid waste treatment under one cover could serve as a useful complement to the conventional solid waste textbooks. The word *reclamation* is substituted for treatment in the title of this publication because all of the systems described in it involve reclamation, whereas the term *treatment* includes disposal with and without reclamation. In anticipation of criticism by purists regarding the use of the phrase "reclamation of solid wastes" in the title, the excuse is offered that the more appropriate wording would be too unwieldy, namely, "Biological Reclamation of the Organic Resources in Solid Wastes."

The sequence and length of the chapters of this book are arranged in order of descending importance of the subject as determined by degree of existing development and extent of practice. Thus, composting is the first subject and receives the longest treatment because the technology of composting is fully developed, and it has been practiced on a municipal scale. There was some question as to the logical positioning of the chapter "Land Disposal of Primary and Secondary Domestic Sewage Sludges." The position directly after "Composting" rather than after "Anaerobic Digestion" was selected because it deals with sewage sludge, whereas the chapter on anaerobic digestion specifically excludes wastewater solids. A second reason is that since the section deals with the utilization of sewage sludge in crop production, the constraints on such utilization are the same as those with the use of compost. Moreover, one of the methods of preparing the sludge for disposal on the land is composting. Reasons for limiting the discussion to land disposal are given in the introduction to the chapter.

The term *reclamation* in the title precludes the inclusion of landfilling, sanitary or otherwise, in the discussion even though biological decomposition has an important role in landfilling. Moreover, the principal activities in landfilling are directed toward containing the wastes and their decomposition products rather than putting them to some useful purpose. Finally, the subject of landfilling is one of those covered most extensively in the usual text on solid waste management.

This author, being of a rather cautious nature, concludes this section with a disclaimer or two: Since the book is not intended to be a handbook, the emphasis is on principles rather than on specific details of individual processes; and, although the hope was for the coverage of the subjects to be as broad as possible, the realization of that hope undoubtedly falls short of the mark, as is the case with so many human enterprises.

1. COMPOSTING

PRINCIPLES

Definition

A definition of *composting* is needed to delineate the subject. It is especially important in a treatise on the use of composting as a solid waste treatment method in order to distinguish it from the ordinary decomposition that is an essential part of the natural order of things. The need becomes apparent upon reading the literature or listening to speakers expatiating upon the various aspects of composting. Some individuals restrict the term to the application of composting as an approach to waste management, whereas others label all decomposition as "composting." The latter use the word *composting* synonymously with "biological decomposition." Clarity and efficiency of reference dictate the use of a definition geared to solid waste management. With this rationale in mind, the following definition was drawn up:

1

Composting is a method of solid waste management whereby the organic component of the solid waste stream is biologically decomposed under controlled conditions to a state in which it can be handled, stored, and/or applied to the land without adversely affecting the environment.

The key words and phrases in the definition, i.e., those that distinguish composting from other processes, are *biologically decomposed, organic component of the solid waste stream,* and *under controlled conditions.* One perhaps could make a case for also regarding the clause which deals with the stability, environmental effects, and safety of the product as being a key phrase. The condition of biological decomposition distinguishes composting from other methods of handling or treating solid wastes, e.g., incineration, sanitary landfilling, pyrolysis, etc. The term *organic component* follows from the condition of biological decomposition. Generally speaking, only material of biological origin (e.g., wood, paper, crop residue, etc.) is directly subject to biological degradation. (The qualification "generally speaking" is inserted in the preceding sentence because a few inorganic materials can be directly acted upon by living organisms, e.g., "mineral" fertilizer; "directly" is used to distinguish the decomposition occurring in the compost process from that resulting from reaction with a microbial by-product, as for example an organic acid.) Perhaps the most important of the key phrases is *under controlled conditions* because it distinguishes composting as a treatment process from the biological decomposition responsible for the recycling of nutrient elements in nature. A very practical reason for inserting *controlled* in the definition is that, without it, certain objectionable practices could be designated as composting. Thus, if controlled were not a condition, an open dump could be labeled as being a compost operation because it involves the biological decomposition of organic matter. Similarly, a mound of unattended animal manure or a mass of untreated rotting cannery wastes could be euphemistically described as composting were it not for their not meeting the requirement of being controlled. The last phrase of the definition, which in effect states that the material be stabilized, is an affirmation of the goal of the process rather than a statement pertaining to the essence of the process. Hence, rather than being definitive in its nature, it is a measure of the termination of the process. Therefore, strictly speaking it is not a key phrase without which the definition would not be complete. Nevertheless, it is inserted in the definition for a practical reason which will become clear later on in this text when the technology of composting is discussed. At this point suffice it to state that it serves as a standard for comparing composting rates.

Classifications

Having defined *composting,* the logical sequence to follow is to classify the various types of composting. Regardless of the insistence of taxonomists to the contrary, classification involves a certain amount of subjectivity and therefore has a little or much, as the case may be, of the arbitrary in it. However, a certain amount of the classification of composting processes can be justified on the grounds of intrinsic differences or similarities. Three bases of classification in composting that have a practical significance are degree of aeration, temperature, and technology. The resulting classes are (1) aerobic vs anaerobic, (2) mesophilic vs thermophilic, and (3) mechanized vs nonmechanized systems. Synonyms for the third classification are *closed* vs *open* composting and *mechanical* vs *windrow* composting.

Aerobic composting is the designation given those compost processes that involve decomposition in the presence of air (i.e., oxygen). Conversely, anaerobic composting implies decomposition in the absence of air. Anaerobic composting is akin to anaerobic digestion used in treating sewage sludges, and indeed the definition fits anaerobic digestion. However, a technological difference does exist between the two. In composting, the wastes are maintained in the "solid" state, whereas in anaerobic digestion the wastes are in slurry form (i.e., slurried), usually at a settleable solids concentration of 5 to 10 percent. Nevertheless, the solid residues of the two processes are the same in terms of condition of the product, namely, a stabilized residue.

Modern compost systems are aerobic — and for important reasons. One of the more important of the reasons is that aerobic processes are not characterized by objectionable odors. Some promoters of aerobic composting are prone to characterize it as being odorless; but by the very nature of the raw material and of the intermediates formed, some odor production is bound to occur. Even the claim of producing no objectionable odors may be inaccurate. An old adage in sanitary engineering is to the effect that there is no such thing as an unobjectionable odor. The reason is that an individual's reaction to an odor is subjective and largely influenced by training and cultural background. For example, the tarry odor of a well-digested sludge may be a stench to a fastidious individual, especially if he is aware that the material is a residue from digesting sewage sludge. On the other hand, to the seasoned sewage plant operator, it may be downright pleasing. Nonetheless, aerobic composting is not accompanied by the foul stench that is the trademark of an unsealed anaerobic composting operation. The reason is that the intermediates formed in aerobic composting are in the oxidized state, whereas those formed in anaerobic composting are in the reduced stage. Aside from the mercaptans, the major sources of stench in anaerobic composting are the short-chain fatty acids, which range from acetic through caproic.

A second and perhaps more important reason for aerobic composting pertains to public health and crop production. After all, an odoriferous anaerobic mass can be "sealed off" from the environment. Public health and crop safety come from the high temperatures that are the natural concomitants of a properly conducted aerobic compost operation. As will be shown later, the temperature in an aerobic pile reaches levels above the thermal death point of most plant and animal pathogens and parasites. These elevated temperatures also are lethal for weed seeds. Anaerobic composting is not characterized by a perceptible rise in temperature. Of course, temperatures could be elevated in an anaerobically composting mass of material, but to do so would involve the expenditure of a considerable amount of energy.

A third reason for aerobic composting is that it is more rapid than anaerobic fermentation. This point may be challenged by champions of anaerobic processes, and perhaps the challenge may have some justification in practical operations. In recent years great progress has been made in accelerating the anaerobic fermentation process through careful design of equipment and operational procedures. However, many organisms that occur in composting and that can rapidly break down refractory compounds are obligate aerobes and therefore obviously could not survive in an anaerobic environment. Moreover, all anaerobic processes call for an eventual aerobic stage to stabilize (i.e., oxidize) the residual intermediates to ensure nuisance-free storage or utilization. The relative speeds and extents of stabilization accomplished in the two approaches can be illustrated by an examination of the contents of a landfill long after it had been completed. Following the first day or two after burial, decomposition of the organic components of landfilled refuse becomes anaerobic because of the exhaustion of the oxygen trapped in the interstices of the buried material. Examination of a completed fill a year after completion of burial will show that many of the organic components can still be identified.

Naturally, aerobic composting is not without its drawbacks. For example, the maintenance of aerobic conditions involves more handling and greater spatial requirements than would be the case with anaerobic composting. On the other hand, the cost of erecting the structures required to contain a nuisance-free anaerobic operation would far exceed that for an aerobic process. Another penalty is the inevitable loss of at least some nitrogen in aerobic composting. The loss is the accompaniment of the high temperature and eventual alkaline conditions reached in an aerobic pile. These two conditions promote the volatilization and loss to the atmosphere of some of the NH_4-N that might be present. As will be seen later, this loss can be minimized, but nevertheless it takes place.

Since modern compost processes are or aim to be aerobic, the remainder of this text, with few exceptions, is concerned with aerobic composting.

Modern composting practice calls for the involvement of mesophilic followed by thermophilic conditions. The reader should note that the term *involvement* rather than *application* is used. The reason for the distinction is that the heat energy responsible for elevating the temperature originates in processes that are internal to the composting mass rather than from the deliberate application of heat from an external source. The classification mesophilic vs thermophilic is based on the temperature range within which the process takes place. For the information of the nonbiologist, *mesophilic* is a term used to designate those organisms for which the optimum temperature is within the range of 8 or 10°C to 45 or 50°C. Organisms having an optimum temperature within the range of 45 or 50°C and higher are regarded as being *thermophilic*. Those having an optimum temperature lower than 4 or 5°C are termed *psychrophiles*. The cut-off point between the various ranges is not sharp, inasmuch as one blends into the next. An important reason for the normal occurrence of a succession of mesophilic followed by thermophilic conditions in composting is that the temperature in an aerobic mass of material inevitably rises to the thermophilic level unless positive measures are taken to prevent it from so doing. Whether or not composting proceeds at a faster pace at higher temperatures is open to question. More is said about this aspect in the discussion on temperature as an environmental factor.

The last major classification, mechanical vs windrow, is based upon the technology involved in the compost operation. *Mechanical* composting involves the use of mechanized, enclosed units equipped to provide control of the major environmental factors. *Open* or *windrow* composting implies stacking the raw material in elongated piles (windrows) and allowing the composting process to proceed therein. Much can be said for or against both approaches. The discussion of the "pro's" and "con's" is reserved for the section on technology.

Nature of the Process and the Implications Thereof

The major implication of composting as a biological process is that factors and requirements peculiar to biological activity in general determine the course of the compost process. This fact brings with it the advantages, limitations, and constraints common to biological treatment processes. The main advantages are a cost lower than that of most physical-chemical systems and a generally lesser production of undesirable by-products. Disadvantages are slowness in comparison to physical-chemical systems and, in some cases, a greater sensitivity to shock loads. Some individuals include another disadvantage, namely, a lower degree of predictability of performance.

As stated before, because composting is a biological process, only organic material of biological origin can be composted. This means that organic compounds of nonliving origin (i.e., organic only in the chemical sense) cannot be composted. Plastics constitute practically all of this type of organic compounds, and indeed plastic material is unaffected by exposure to the compost process — unless it is formulated of a biodegradable resin.

As a biological process, a microbial population is essential to the functioning of the compost process. Moreover, it follows then that the size and composition of the microbial population determines the rate and extent of the compost activity. The microbial population must consist of microorganisms capable of using as a nutrient substrate the material to be composted. In other words, the material must be of a type that is assimilable by the microbes. The rate of the process is that of the activity of the microbial population active within it. A corollary is that a definite upper limit exists beyond which size of microbial population and sophistication of equipment can do nothing to further speed up the process. Ultimately, i.e., with all environmental factors at an optimum level, the rate of bacterial activity is limited by the genetic make-up of the individual microorganisms. This fact has a practical significance with respect to compost technology in that it is useless to spend money on equipment designed to accelerate the process beyond that permitted by the genetic make-up of the microbial population. A similar reasoning applies to the limitation on the capacity of the system that arises from its biological nature. The more the material to be composted and the speedier it is to be done, the greater must be the size of the active microbial population. A corollary is that the more diverse the material, the more diverse should be the microbial population.

Yet another result of the essentially biological nature of the compost process is that all environmental factors that affect any biological activity also affect for good or ill the composting process. The aim of compost technology, therefore, should be to maintain these conditions at an optimum level. However, because of the economic uncertainty of large-scale compost operations, it may not always be economically feasible to use the equipment and procedures needed to provide optimum conditions. The recourse then is to compromise by providing an environment favorable to the extent allowed by economic necessity.

Microbiology

■ **Principal Groups and Their Roles:** Before discussing in detail the environmental factors and their relation to the compost process, it would be well to deal with microbiology of the process; inasmuch as we have seen, the success or failure of its outcome depends ultimately upon microbiology. The discussion on microbiology can justifiably cover a wide range of subjects, extending from a listing of the names of the microorganisms involved to the advisability of using inoculums. Somewhere in between is room for an evaluation of the utility of making a comprehensive identification of all of the microbes involved. In this presentation, the dissertation on the "pro's" and "con's" of inoculums and of the utility (or futility) of attempting to identify and quantify all of the organisms in a "typical" composting pile is preceded by a listing of those organisms already known and a cursory examination of their respective roles in the process.

The three main classes of microorganisms to be encountered are bacteria, actinomycetes, and fungi. The actinomycetes have been regarded as being "higher" forms of bacteria. However, according to Hesseltine[1] the actinomycetes are characterized by enough properties of a type to warrant placing them in a phylogenetic line separate from the bacteria and the fungi, but more closely related to the fungi. The latter judgment was based on the many fungal-like characteristics of the actinomycetes. Regardless, the distinct role played by the group in composting warrants treating them as a separate group in this text. Reports in the literature on attempts to identify bacteria involved in the compost process are not numerous.[2] In this author's research, species of *Pseudomonas* were isolated in the few efforts made to identify bacterial types. The identification of pseudomonads as a part of the bacterial population in composting led some of the early proponents of inoculums to claim that the introduction of the organisms into composting material would provide an added source of nitrogen. They had heard that some pseudomonads could fix atmospheric nitrogen. Their hopes were far-fetched. Even if the organisms were capable of fixing nitrogen, conditions in a typical compost operation would not be conducive to nitrogen fixation.

The lack of reported studies on bacterial identification does not imply a minor role for the bacteria in the process. If this author were so rash as to ascribe a numerical value to their portion of the total activity, he would intuitively hazard an estimate of 80 to 90 percent. The bacteria certainly are responsible for the initial breakdown of the organic material and for a large part of the heat energy released into the composting mass.[2] An extensive role would be expected of bacteria because of the nature of the material being acted upon. Even with a relatively homogeneous mass, such as a given crop residue or animal manure, a wide variety of compounds must undergo biological decomposition. These compounds range from complex proteins and carbohydrates to simple sugars and amino acids. While it is true that certain classes of fungi and actinomycetes collectively can utilize a fairly wide array of substrates as a nutrient source, they do not appear in visibly appreciable numbers until the process has been well advanced.

Species of the actinomycete genera *Micromonospora, Streptomyces,* and *Actinomyces* can regularly be found in composting material. The three genera may be distinguished from one another in a nutrient medium by morphological characteristics, such as the nature and formation of substrate mycelium and aerial mycelium (if any), the formation of conidia or of spores, and the size and shape of the sporophores. Attempts made to classify to species level the actinomycetes in a compost pile will prove to be arduous and tedious. Should one embark upon the task, he or she would be well advised to use the excellent textbooks written by Waksman, one of the latest of which is *The Actinomycetes: A Summary of Current Knowledge.*[3]

The presence of actinomycetes in a composting mass can be readily detected visually and olfactorily. Under favorable conditions the composting material begins to acquire a faintly earthy odor after five or six days have elapsed. The odor becomes more pronounced as time progresses. In his studies, this author noted that material being composted in a room in which the temperature was held at 50 to 55°C acquired an earthy odor within a few days that promptly was transformed into a rank, fishy smell upon removal to a cooler environment. The fishy odor may have been produced by a thermophilic actinomycete normally not encountered when the compost operation is carried on at an ambient temperature within the mesophilic range. The presence of actinomycetes does not become visually detectable (i.e., by the unaided eye) until the course of the process nears its end. When they do become apparent, they appear as a blue-gray to light green powdery to somewhat filamentous layer in the outer 4 to 6 in. (10 to 15 cm) of the pile. At times, circular colonies can be detected on bits of cardboard or on fruit wrappers. Such colonies expand slowly, leaving a correspondingly widening void in the center of the colony. However, the bluish-green zone does not consist solely of actinomycetes: it also has a more or less large concentration of fungi intermingled with the actinomycetes. Obviously, this visual evidence of the presence of actinomycetes does not occur in mechanized composting because of the mixing action that takes place in the digester.

The actinomycetes exert their greatest effect on the cellulosic and, to some extent, the lignaceous components of the composting mass. Thus, paper, which may have been but little affected previously, appears to be rapidly disintegrated upon the appearance of the actinomycetes. This is due to the fact that most actinomycetes can utilize a relatively wide array of compounds. Consequently, as sources of energy they can use organic acids, sugars, starches, hemicelluloses, cellulose, proteins, polypeptides, amino acids, and a large number of other organic compounds. According to Waksman,[3] some species even can decompose lignin, tannin, and natural rubber to some extent.

Interestingly, this author could detect neither a gross olfactory nor a visible manifestation of the presence of actinomycetes in the three municipal compost operations visited by him in Mexico. Composts in various stages of maturity were examined, and at no time could an earthy odor be detected. Nevertheless, the finished compost had the physical and chemical properties characteristic of a compost in which actinomycetes had been more visibly present. The reason for the apparent absence of actinomycetes in the Mexican compost piles may have been a relatively small amount of paper content and hence a lesser amount of available cellulose. Whereas the garbage (wastes from food preparation and other putrescible wastes) averages around 10 percent in United States refuse, that in Mexican refuse may be as much as 50 percent.

Finally, it should be emphasized that the lack of a visible layer of actinomycetes does not preclude their being present in significant numbers, inasmuch as they may have been dispersed throughout the composting mass. However, this does not seem to be likely in view of the fact that internal temperatures were above those suitable for the organisms, namely, in the 55 to 65°C range.

The presence of fungi becomes obvious at about the same time as that of the actinomycetes, and in the same zone. Indeed, as stated earlier, they grow intermingled with the actinomycetes. More types of fungi have been identified in the compost process than has been the case with bacteria and actinomycetes. Contributing factors to this situation have been the relative ease of identifying the fungi and the search for new types by specialists in thermophilic fungi. Names of fungal groups encountered by this author in the literature or identified by him in his research are *Chaetomium thermophile, Humicola languinosa, Talaromyces dupontii, Aspergillus fumigatus, Mycogone nigra, Botryosporium* sp., and *Stactybotrys* sp. Undoubtedly, other species or groups have been reported and can be found in composting material. That a large number of isolates can be made from a single operation was demonstrated by Kane and Mullins.[4] They managed to isolate 304 unifungal cultures from one batch of compost processed in a mechanical-type composter. Of the 304 isolates, 120 belonged to the genus *Mucor,* 94 to the genus *Aspergillus,* 78 to the genus *Humicola,* six to the genus *Dactylomyces,* two to the genus *Torula,* one to the genus *Chaetomium,* and three unidentified. Of course, as Kane and Mullins are careful to point out, the number of isolates does not necessarily indicate the actual number of various organisms nor their relative activity in the process. Not surprisingly, they were able to isolate thermophiles at all stages of the process, i.e., even before thermophilic conditions had set in. This fact is not unexpected because obviously the spores or the organisms themselves had to be in the raw material or had to have been carried to the material by way of air currents. Some concern may be felt because of the detection of *Aspergillus fumigatus* in composting material, inasmuch as the group has been regarded as being potentially pathogenic. However, no human infections by the organisms due to exposure to compost have as yet been reported.

The effect of fungal activity on the compost process is closely akin to that of the actinomycetes in that they have the same versatility with respect to substrate utilization as a nutrient source. Perhaps this can be regarded as another bit of evidence in favor of the close relationship between fungi and actinomycetes. The similarity of substrate utilization, the coincidence of their appearance, and the comparability of their gross morphology make it difficult to assess the individual effectiveness of each of the two groups of microorganisms. It follows therefore that the resolution of the question would be only of academic interest.

An intriguing question is the identity of the factor or factors that limit the gross appearance of the fungi and actinomycetes to the outer 10 to 15 cm (4 to 6 in.) layer of a pile. The answer to this question has a practical significance in mechanical composting in which, by the very nature of the mechanized process, layer formation could not occur. The bearing is in the form of a retardation if inhibiting conditions of the kinds found beneath the 10- to 15-cm outer layer were to prevail throughout the continuously mixed contents of a digester. The consequences are obvious, and the most apparent would be a slowing of the decomposition of paper. Two limiting factors can be suggested: an excessively high temperature and an inadequate oxygen supply. This author made no attempt to resolve the question of which one of the two factors or whether both constitute the limiting condition.

Evidence in favor of inadequate aeration rests on the fact that a large number of the actinomycetes and most of the fungi are obligate aerobes. In a stationary pile, completely aerobic conditions probably prevail only in the outer layer. If oxygen were the limiting factor, then the two groups could flourish in a mechanical digester provided it were adequately aerated. Evidence in favor of high temperature being limiting is that the temperature in the interior of the pile at the onset of the mass appearance of the two groups of organisms may be as high as 65 to 70°C. The zone of high temperature extends to within 15 or 17.5 cm (6 or 7 in.) of the exterior of the pile. These temperatures probably exceed the thermal deathpoint of most representatives of the two groups. The temperature concept with respect to fungi species is borne out by Finstein and Morris,[2] Kane and Mullins,[4] and

Gray.[5] They all report failure of fungal isolates from compost to grow at temperatures of 60°C or higher. Another condition for the development of the layer was a sufficient time interval between turnings, i.e., mixing the contents of the pile. By the time of the development of the layer in the author's studies, the turning frequency had been reduced to once every four or five days. Therefore, growth rate (and thus, indirectly, frequency of turning) could be regarded as a contributing factor to layer formation — but not to limitation of the depth of the layer.

■ **Utility of Identification to Process Design**: The subject of microbiology in composting brings up what may be a controversial question, namely, would the exercise of identifying all of the microbes involved, if it could be accomplished, have any practical utility? The position one holds on the subject usually depends upon his or her professional background. Generally, the microbiologist and perhaps the chemical engineer would have no doubt as to the imminently practical utility of such an attempt. On the other hand, the sanitary engineer would take a less sanguine view regarding the utility.

Before taking a stance on the subject, one should consider the following arguments for and against identification. (The arguments are conditioned by the assumption that the undertaking to identify would involve the expenditure of an intense and perhaps prolonged effort to make the necessary isolations.) The discussion of the "pro's" and "con's" begins with a consideration of the motivation or reasons for identifying the various microbial populations. Obviously, a major motive would be to improve the process. A secondary motive would be academic in nature, namely, to add to the existing fund of knowledge and thereby lead to a further understanding of the interrelationship between organisms and a broader knowledge of the constitutive organisms. The improvement in the process would result from the determination of the qualitative and quantitative aspects of the environmental factors conducive to the maximum desirable activity of each of the microorganisms, the collective activities of which constitute the compost process.

It is quite apparent that if growth requirements of an organism were known, equipment and operational procedures could be designed to provide for those requirements. But to determine the requirements, it is necessary to identify the organisms for which they are to be provided. In practice it would be a waste of effort to determine the requirements and subsequently design to meet them for each and every group of microbes that may happen to be present in composting material. Rather, the aim should be to design only for those organisms that play a significant part in the compost process. In fact, it could even be disadvantageous to design for certain groups of organisms, since they might flourish at the expense of the more useful ones. Hence, it is necessary not only to identify the microorganisms present, but also to determine the nature and extent of their contribution to the process as a whole.

● *Uncertainties in the Determination of Roles* — The problems to be encountered in isolating key microorganisms in composting are many, albeit common to all isolations and identifications of organisms from "natural" situations. Nevertheless, they are amenable to conventional microbiological techniques. The major difficulty, or rather uncertainty, arises from the fact that to ensure the isolation of all of the significant groups, it is necessary to provide a medium that meets the nutritional and other requirements of each of those groups. In other words, it must be assured that no particular group should escape detection because of a failure to provide a medium on which its members could grow.

The problem of a suitable medium has a bearing on another difficulty, namely, that of determining the true extent of the abundance of a given type of organisms. It is very possible that the medium and cultural conditions used in making the isolations might constitute an enrichment situation for a group present in very small numbers (e.g., "group A"). Of course, this possibility could be countered by a proper streaking and plating technique, but then chances would be favorable for every individual organism to survive and multiply enough to be registered in the plate count. On the other hand, if the isolation medium were not suitable, or only slightly so, for a given group present in large numbers ("group B"), none or very few would survive and multiply to an extent sufficient to be "countable." The false conclusion to be reached solely on the basis of the two situations would be that "group A" was present in great abundance in a given pile, whereas "group B" was not present or only sparsely so. At times the problem of a suitable medium can be solved at least in part by using an "extract" of the material to be analyzed. The "extract" can be (1) a finely dispersed suspension of the material, (2) the liquid remaining after the suspended material has been removed, or (3) the filtrate obtained by filtering a suspension that has been heated. Generally speaking, however, there is no guarantee that the relative abundance of the microorganisms can be determined with any great degree of certainty.

The next problem, that of assessing the significance of a particular group of microbes, probably is more difficult to solve. The exception is when the significance is quite obvious. An example of the latter is that of the actinomycetes and fungi. The development of the bluish-green layer, described in a preceding paragraph, coincidental with the disappearance of paper readily leads one to the palpable conclusion that at least in windrowed composting material actinomycetes and fungi have a significant role in decomposing paper — and hence cellulose. The role of these two groups of microbes, however, is not so apparent in a mechanical compost operation because of the absence of "layering."

Abundance is not necessarily a measure of significance; although if abundance were real and not an artifact of the analytical conditions, chances are that it would be a concomitant of significance. The reason is that abundance in most cases would be due to the ability of the organisms to use the substrate as a nutrient source, and the act of utilization as a nutrient source constitutes the decomposition of the substrate. A qualification is that the conversion efficiency of some microorganisms might be low, and therefore more substrate would be needed to produce a given amount of cellular material (i.e., cells) than would be the case with other microorganisms. Consequently, a group of such low efficiency microorganisms present in smaller numbers could be more significant in terms of substrate degradation than an efficient organism present in large numbers, especially if nutrient availability were not a limiting factor.

Once a microorganism is isolated, the determination of its optimum cultural conditions is not an unduly difficult task, except with a type that has an undetermined growth factor requirement or is limited with respect to the range of substrates that it can assimilate. Well-established procedures are available for relating growth characteristics to environmental conditions. (Here, environment is taken in the context of nutritional as well as physical and chemical.) The problem is that the conventional procedures call for the determination of optimum conditions with the organism growing as a pure culture, i.e. monoculture. It is difficult to answer with certainty whether or not a set of conditions proven to be optimum for a group of microorganisms grown in monoculture would remain optimum for that group when it is but one of many groups, as is the case in the compost process. Another problem would be that of reconciling "conflicting" optimum conditions. For example, the optimum temperature for one set of microorganisms might be 35 to 40°C and 30 to 35°C for another.

Finally, granted that the significant groups in a mass of composting material are identified and their growth requirements evaluated, the next question concerns the universality of the findings. The answers to the following two questions determine the degree of universality: Would the microbes found in one type of system (e.g., windrow) also occur in other systems (e.g., enclosed or mechanized)? Would the organisms found to be of significance in one type of material (e.g., manure) also be of importance in another (e.g., refuse)? The degree of universality of any findings is a measure of their utility and, hence, of the justification in expending the effort needed to make them.

The fact that composting is an integration of the activities of a wide variety of microorganisms is the most weighty argument against the utility of trying to determine an optimum set of operating conditions through a piecemeal approach, that is, through a course of isolation, identification, kinetic and physiological studies, etc. The more logical approach would be to determine the optimum design for equipment and operational procedures by working with the integrated system and not with each of its components in isolation. In other words, it would be more sensible to directly measure the effect of levels of key environmental factors on the performance of the composting material *en masse*. Design factors thus determined would suit the process as a whole and not merely a single group of organisms. Even if it should turn out that a particular type of organism exerts an influence greater than that of any other on the rapidity or efficiency of a given system, the utilization of the integrated approach, by its very nature, results in a value for each factor consonant with the significance of the major as well as of the lesser microorganisms. It is true that such an approach involves some trial and error, but not significantly more than occurs in most research.

A practical point to keep in mind is that, regardless of which approach is followed in ascertaining the optimum design factors, costs of implementing them most likely will be excessively high in terms of the monetary value of the product and of the service supplied, as has been demonstrated by the history of mechanized composting in the United States. However, by working with the integrated system it is possible to arrive at a compromise level at which the composting process could be maintained at a satisfactory level and pace.

Inoculums

The subject of inoculums is an appropriate one for the section on microbiology. Inoculums and inoculation are used here in the microbiological sense of introducing a relatively minute aliquot of a suspension of bacteria or of other microorganisms into a much larger volume of medium (e.g., 500 ml of a 10^6 bacteria/ml suspension into 1 metric ton of refuse). This usage differs from the "mass inoculation" often used in sanitary and chemical engineering. Mass inoculation involves the introduction of a volume of microbial culture sufficiently large to significantly affect the composition of the receiving material and to supply a sizeable fraction of the eventual peak microbial population.

If an inoculum were to be useful in composting it would have to (1) supply a needed type of microbe not already present in the material to be composted, (2) augment a less than adequate population of microbes, or (3) introduce a group of microbes supposedly more effective than any indigenous to the raw material. The first condition obviously does not prevail in wastes usually encountered because unless sterilized or otherwise especially treated they inevitably decompose at a rate limited solely by environmental conditions. Moreover, the absence of a specifically essential microbe from a specific mass of material must be demonstrated before the microbe could be added with any practical effect. To demonstrate such a lack would be a well nigh impossible task, as was explained in the preceding paragraphs. The problem is made complex by the fact that the needed types vary with the composition of the raw material. Where special types of wastes (e.g., sawdust) are composted without being mixed with other wastes, the slowness of the material to decompose is more a function of imbalance or even complete deficiency of needed nutrients or of resistance to bacterial invasion.

The arguments advanced in the preceding paragraph also apply to the second condition, namely, augmentation of an inadequate bacterial population. The third condition, introduction of a supposedly more effective type of bacteria, is the one commonly advanced by individuals engaged in the business of selling inoculums. Several facts combine to cast doubt on the utility or even the possibility of speeding up the compost process simply through the introduction of a theoretically more active organism. Despite the existence of these facts, new inoculums are introduced each year with the claim that the compost process can be accelerated materially with the use of the "especially developed" microorganisms.

Several reasons can be advanced against the introduction of supposedly more active microbes. Before giving the reasons, it should be pointed out that first it must be demonstrated that such a special strain has been developed. This author has yet to encounter such a strain, nor has a satisfactory proof of the existence of one been reported in the scientific and technical literature. But in the unlikely event that such an inoculum could indeed be developed, an argument can be made regarding the futility of using it. The first fact militating against the possibility of acceleration through introduction of a group not already present is that the introduced organisms are at a disadvantage with respect to the indigenous population. The latter is adapted to the conditions prevailing in the waste and hence have a competitive advantage. By the time the introduced microbes have become adapted, if that could be done, the indigenous population will have increased to the limits permitted by the existing supply of nutrients. Consequently, the presence of the added microbes would be superfluous.

A second reason arises from the nature of the compost process. In composting, as in the decomposition of any complex substance, the breakdown is a dynamic process accomplished by a succession of microorganisms, each group of which reaches its peak population at the time conditions have become optimum for it. The optimum conditions usually are nutritional in nature. Since other conditions (e.g., moisture content and aeration) usually are kept constant, the principal change other than nutrient is temperature. This latter is important in the transition from predominantly mesophilic to thermophilic populations. The nutritional barriers normally are those of degree of complexity and, hence, availability of the elements (carbon, nitrogen, phosphorus, etc.) essential to a given type of microbe. In nature, some microorganisms are able to utilize complex molecules as sources of essential nutrients, whereas others can assimilate only rather simple compounds. Others even may be restricted either to inorganic (e.g., NH_4-N, NO_3-N) or to organic compounds (e.g., amino acid-nitrogen).

The net result of these requirements and limitations is a successive appearance of groups of microbes adjusted to a stepwise reduction of complex substances to simpler compounds, much like the ecological succession that occurs with higher plants in nature, but obviously here at an exceedingly faster pace. Certainly it would be of little use to introduce a given microorganism before its particular "niche" was ready. Populations of the indigenous microbes begin to expand as nutrients become available to them and at a rate proportional to the availability of the nutrients.

Types of organisms that might bypass or rather encompass this succession are the fungi, inasmuch as certain groups can "attack" substances as complex as paper and wood. However, there must be some limiting factor or factors for the fungi in the first stages of the compost process because they do not appear in full density until the process is well advanced, even though their presence can be demonstrated throughout the course of the process.

Another factor to keep in mind is that strains of microbes continuously subcultured under laboratory conditions tend to become attenuated. For instance, the infective ability of pathogenic bacteria is markedly diminished through repeated subculturing. Therefore, granted that specific microbes demonstrated as being superiorly active are isolated, in time the stock culture gradually loses its pristine characteristics and thus becomes unable to compete with indigenous populations.

In recent years many microbiologists, overimpressed by the potential of influencing biological treatment systems through the induction of mutation, propose that approach in composting. Making such an approach unfeasible are all of the factors discussed in the preceding paragraphs.

In the unlikely event that an inoculum may be required, two excellent sources can be used. One is a shovelful (or more, if need be) of rich garden loam or of partially decomposed horse manure. These materials contain all the microorganisms — from bacteria to and including fungi — needed to decompose (compost) any material. Another source is composted material. The latter brings up the matter of mass inoculation through the introduction of a portion of the completed compost into the incoming wastes, i.e., recirculation of a part of the product. There are conflicting reports on the utility of such a course. In his studies, this author could perceive no significant acceleration of the process through such a practice. However, if enough were recycled, it could improve the texture of a raw material, if that should be needed.

A final note of caution: In evaluating the utility of a particular inoculum, care should be taken to provide "controls." The control pile(s) (not receiving the inoculum) should be subject to *all* of the conditions imposed upon the inoculated pile. For example, if the inoculum changes the pH of the treated pile, that of the control pile should be adjusted to the same level. The same applies to changes in the level of nitrogen and other nutrients. In other words, the *sole* difference between the inoculated and uninoculated piles should be that one is inoculated and the other is not. Both piles should be set up and managed simultaneously. While these precautions seem obvious, a review of reports of so-called improvement through inoculation invariably has shown that either no controls were set up or that the conditions applied to the controls differed from those imposed upon the inoculated piles.

The final argument, and perhaps the most convincing one, against the usefulness of inoculums in routine composting is in the negative results reported by most researchers in objective studies on inoculums. Such studies have been numerous. The conclusion of the researchers has been that if the addition of microbial "starters" or "promoters" contributes anything to facilitate the compost process, it is so minute as to be undetectable.[2,6-9]

Environmental Factors

The environmental factors can be grouped under the headings "physical," "chemical" and "nutritional." Broadly speaking, certain nutritional factors are both chemical and nutritional. The major nutritional factors are concentration and availability of nitrogen, phosphorus, potassium, and carbon. A chemical factor is concentration of oxygen. Physical factors are temperature, moisture content, and degree of mixing of the composting mass.

■ **Substrate (Nutritional):** Substrate may be treated as an environmental factor because it is extrinsic to the microbial population and, as such, exercises an influence on the extent and rate of bacterial activity.

In composting, as with any biological process, the physical and chemical nature of the substrate is a key factor in determining the course and rate of the process. Essentially, it is the degree and ease of availability of nutrients to the various microorganisms, as well as the quantity and balance of the nutrients. Pertinent physical characteristics of the substrate are primarily related to particle size and the moisture content of the mass of the material. Particle size is important because it determines the surface area per unit of mass exposed to bacterial attack. Chemical characteristics of importance are primarily those related to molecular size and complexity and nature, as well as to the elemental make-up of the molecules.

Complexity and nature of molecular structure of the substrate are especially important because they determine the assimilability (i.e., vulnerability to attack) of the nutrients in the substrate by the various microorganisms. Obviously, if the substrate cannot be assimilated by any of the organisms present, it will not be composted. The capacity of a microbe to assimilate a given substrate depends upon its ability to synthesize the enzymes involved in breaking down complex compounds into intermediate compounds or into an element that can be utilized by the microorganism in its metabolism and synthesis of new cellular material. If the microbes collectively lack the required enzymes, the substrate remains unscathed. The more complex the compound, the more extensive and comprehensive is the enzyme system required. Some molecular structures are of a kind that can be assimilated by only a few groups of microorganisms. Translated into practical application, this limitation means that substances consisting mainly of cellulose (e.g., paper), of lignin (wood), or of molecules having a ring structure (e.g., aromatics) break down more slowly than do highly proteinaceous materials (meat scraps, fresh vegetable trimmings, garbage, etc.). The significance of nutrient availability and balance is explained in the section that follows.

■ **Nutrient Balance:** Two factors enter into the determination of a suitable nutrient balance: (1) the elemental composition of the microbial cell mass and (2) microbial metabolism. For reproduction and consequent decomposition to take place, all microorganisms must have access to a supply of the elements of which their cellular matter is composed. Other elements are required which enter into the metabolic activities of the organisms by serving as an energy source or as an enzyme constituent. These latter do not necessarily increase the cellular mass of the organisms. The respective amounts required from the elemental make-up of the substrate molecules determine their nutritional utility to the microbes. The more abundant the elements of nutritional significance in a substrate to microbes, the greater will be the number of microorganisms supported by it, and hence the more extensively and rapidly will it be composted.

Almost all elements are utilized to some extent by microbes, and some are essential to their survival. The relative amount required of each element varies. Those needed in large amounts (e.g., grams or mg/liter) are termed *macronutrients;* those in minute amounts (e.g., micrograms/liter) are designated as *micro-nutrients* or trace elements. The principal macronutrients are carbon (C), nitrogen (N), phosphorus (P), and potassium (K). Their rating as macronutrients is dictated by their abundance in the elemental composition of microbial cellular material. About 50 percent of the cell mass consists of carbon, and from 2 or 3 to 8 percent is nitrogen. The ratio depends upon the type and cultural conditions of the microorganisms. Potassium is present in only a fraction of a percent. Hydrogen and oxygen constitute a large percentage of the cellular mass in the form of water and as a part of the cellular material. The major part of the molecular structure of the cell mass has the basic formulation (CH_2O). Trace elements, while needed in only minute, trace quantities and are even toxic above those traces, nevertheless are essential to the survival and multiplication of the microorganisms. Some, or perhaps most, of the trace elements constitute an important component of the enzyme molecules.

The respective amounts of elements required vary at a somewhat constant ratio to each other. In other words, a balance is struck. The need for a balanced supply of nutrients becomes essential when one or more of the nutrient elements is present in less than the concentration at which bacterial growth is limited by a factor other than nutritional. In practical operations this balance is especially important as far as macronutrients are concerned.

• *Carbon-Nitrogen Ratio* — One of the more important aspects of the total nutrient balance is the ratio of carbon to nitrogen. Not unexpectedly, the ratio commonly is termed the carbon-nitrogen ratio (C/N). A material having, for instance, 30 times as much carbon as nitrogen is said to have a C/N ratio of 30/1, or simply a C/N of 30. Although other desirable nutrient ratios exist (e.g., nitrogen to phosphorus), they rarely need to be considered in composting wastes. Therefore, the carbon-nitrogen ratio remains the most important balance in composting. Consequently, the C/N ratio is very important in assessing the suitability of a given waste as a substrate for composting. Living organisms require available carbon as a source of energy and need nitrogen to synthesize protoplasm. Inasmuch as the efficiency of living organisms necessarily is less than 100 percent, more carbon than nitrogen is needed. However, if the excess of carbon over nitrogen is too great, biological activity diminishes. In a composting operation the manifestation could be an excessively long time required to complete the process, because several life cycles of organisms are required to reduce the C/N ratio to a more suitable level. The lowering of the C/N ratio is brought about by the fact that two-thirds of the carbon consumed is given off as CO_2, while the other third is combined with nitrogen in the living cell. Upon the death of the microbe, fixed nitrogen and carbon again become available, but their utilization once more necessitates the burning of a fraction of the carbon to CO_2. Thus the amount of carbon is reduced by way of partial conversion to CO_2, while nitrogen continues to be recycled.

The possibility also exists for the energy source to be less than that required for converting available nitrogen into protoplasm. In such an event the organisms make full use of available carbon and eliminate the excess nitrogen as ammonia. If the excess of nitrogen in a decomposing mass is too great, i.e., too low a C/N ratio, ammonia may be formed in amounts sufficient to be toxic to the microbial population. If the material had been incorporated into the ground, the NH_3 (ammonia) would be toxic to roots of higher plants with which it came in contact. In an anaerobic situation, excess carbon may lead to acid formation and eventual killing off of the microorganisms. Less drastically, an imbalance results in an inhibition of microbial activity.

The optimum C/N ratio is to some extent a function of the nature of the material of the wastes, especially of the carbonaceous component. If the carbon is bound in compounds broken down with difficulty by biological attack, its carbon accordingly would become only slowly available to the microbes. Therefore, as far as the microorganisms are concerned, only a part of the carbon is available at any one time — perhaps in the entire amount they could possibly use at that time. Compounds of this sort are the resistant ones mentioned earlier, chiefly lignin, some aromatics, and some physical forms of cellulose. Consequently, if a significant portion of the carbon is in the form of such compounds, the permissible C/N ratio could be higher. Except for the small amount of nitrogen in keratin and similar resistant nitrogenous compounds, most of the nitrogen in wastes is readily available.

Consequently, the maximum end of the permissible C/N ratio is rarely lowered by reason of difficultly available nitrogen.

Results of the cumulative experience of researchers in composting over the past couple of decades lead to the conclusion that a C/N ratio of about 25 or 30 parts of carbon to 1 of nitrogen (i.e., C/N 25/1 or 30/1) is optimum for most types of wastes, especially municipal refuse. This is about what one would expect since living organisms utilize about 30 parts of carbon for each part of nitrogen. If the wastes have a large percentage of woody material or newsprint, the permissible ratio can be as high as 35/1, and perhaps 40/1. As indicated before, the only observable penalty for having a C/N ratio lower than 20/1 is loss of nitrogen. Conditions in an actively composting pile are conducive to the volatilization of any NH_4-N that might be present in the composting material. These conditions are elevated temperature and a pH above 7 (usually 8 to 9). They are the very conditions applied in making laboratory determinations of NH_4-N concentration. The loss occurs primarily when the material is turned in windrow composting or tumbled in a digester. Generally, the outer layer of material in an undisturbed windrow prevents the ammonia from escaping from the pile.

The formation of ammonia is the result of the failure of the microbes to use all of the nitrogen released through their breaking down the nitrogenous compounds in the synthesis of new cellular material. The presence of excess nitrogen in the form of NH_4^+ or NH_3 can be traced to the microbial metabolism of protein. Proteins are collections of amino acids, and the amino group $(-NH_4)$ is a characteristic of amino acids. In attacking a protein, the microbes utilize enzymes to break down the protein molecule to its constituent amino acids. Generally, the amino acids will be subjected to further biochemical reactions and thereby become constituents of the protoplasm of microorganisms assimilating it. But if not enough carbonaceous compounds are present to accommodate the amino acids or their intermediates in the synthesis of new cellular material, the excess amino acids will be deaminated. The amino group then is removed and becomes ammonia. An example of deamination is shown in the following reaction:

$$CH_3 CHNH_2 COOH + 1/2 O_2 \xrightarrow[\text{amino acid oxidase}]{} CH_3 COCOOH + NH_3$$

It should be kept in mind that the foregoing explanation of the conversion of excess nitrogen to ammonia is an oversimplification. For a more complete description the reader should consult a textbook on bacteriology, e.g., *Microbiology* by Pelczar and Reid.[10]

The nitrogen concentration of a sample of the material to be composted can be determined by the usual Kjeldahl method. If access to equipment for carbon analysis is not available, the carbon content can be roughly estimated according to a formula developed by New Zealand researchers[11] in the 1960s. The formula is as follows:

$$\% \text{ carbon} = \frac{100 - \% \text{ ash}}{1.8}$$

In the University of California studies it was found that the results of the method approximated the more accurate laboratory determination of carbon within 2 to 10 percent.

An indication of the C/N ratio of various substances can be gained from the list of C/N ratios in Table 1. It is important that the reader keep in mind that the numerical values are not constant, i.e., universally applicable. Even material from the same source varies in composition from day to day. For instance, the C/N ratio of manure is a function of the dietary intake of the animal. That of municipal refuse is strongly influenced by myriad factors, not the least of which are region of the country, size, and industrial nature of the municipality. Despite the potential variations, the numbers do help in assessing the potential compostability of a given raw material or collection of raw materials. They also can be used in deciding which material to add as a nitrogen source to bring down an uncomfortably high C/N ratio or, conversely, the carbon source needed to elevate an excessively low C/N ratio.

In the 1950s and earlier, municipal refuse typically had a C/N ratio within a range suitable for composting; but as time progressed, the ratio steadily rose until at present it is on the order of 60/1.[12] This latter is considerably too high a ratio to be conducive to satisfactory composting, at least at a rate consonant

TABLE 1. C/N RATIO OF VARIOUS WASTES

Material	Nitrogen	C/N Ratio
Night soil	5.5–6.5	6–10
Urine	15–18	0.8
Blood	10–14	3.0
Animal tankage		4.1
Cow manure	1.7	18
Poultry manure	6.3	15
Sheep manure	3.8	
Pig manure	3.8	
Horse manure	2.3	25
Raw sewage sludge	4–7	11
Digested sewage sludge	2–4	
Activated sludge	5	6
Grass clippings	3–6	12–15
Nonlegume		
vegetable wastes	2.5–4	11–12
Mixed grasses	214	19
Potato tops	1.5	25
Straw, wheat	0.3–0.5	128–150
Straw, oats	1.1	48
Sawdust	0.1	200–500

with the efficient use of equipment and space and the production of a high-grade product. However, a point to remember concerning municipal refuse is that as resource recovery becomes more widely practiced, increasing amounts of its carbonaceous fraction will be removed in the form of paper. The result will be that the organic residue from such operations will have a much lower C/N ratio than that of unprocessed refuse. For example, processing refuse through a separation process developed at the University of California results in an organic residue characterized by a C/N on the order of 16 to 20/1. Another way of lowering the C/N of municipal refuse is to add sewage sludge to it. This course of action is receiving increasingly favorable attention, so much so that it is made the subject of a later chapter in this book.

● *Trace Element and Enzyme Requirements* — Generally, elements other than the macronutrients are present in most wastes in an abundance sufficient to permit satisfactory composting without the need for further additions.

The matter of additives to compensate for inadequacies with respect to nutrients leads to that of the utility of supposedly accelerating the compost through the introduction of appropriate enzymes. Theoretically, the addition of enzymes should lead to some acceleration. The difficulty is that it is first necessary to identify the enzymes, assess their role in the compost process, and then find a way to produce them within a cost constraint dictated by the monetary value of the compost product or commensurate with the savings resulting from any acceleration that may be brought about. In practice, the addition of enzymes most likely would be an unnecessary step inasmuch as the microbial population synthesizes the enzymes needed for their activity and nutrient assimilation at a rate parallel with the limiting influence of the environmental complex. Thus far, no conclusive or even significantly supportive evidence for the utility of enzyme addition can be found in the literature.

- ■ **Physical Conditions:**
- ● *Temperature* — In the section on classification, the effect of temperature on the compost process was mentioned somewhat cursorily with respect to its effect on rate of composting. The question is not so much one of the beneficial effect of increase in temperature within the range of either the mesophilic or the thermophilic zones as it is for the overall advantages of carrying on the process at mesophilic as contrasted to thermophilic temperatures. An optimum temperature range for the various groups of microorganisms exists within the major temperature zones, i.e., for the mesophiles when the temperature of the mass is within the mesophilic range and for the thermophiles when it is within the thermophilic range. Thus, the optimum for the bacterium *Pseudomonoas delphinii,* a mesophile, is 25°C; whereas for *Clostridium acetobutylicum,* another mesophile, it is 37°C. Since the composting process represents the integrated activity of a number of different microbes, chances of the temperature being optimum at any single instant for every individual group of the microorganisms present would be nil. However, at peak activity of the pile as a whole, the chances of the temperature being within the vicinity of the optimum would be good. The high degree of

activity is indicative of a satisfactory temperature for most of the microbes. Consequently, it may be concluded that the optimum temperature for the process as a whole is an integration of or, perhaps better expressed, a compromise between the optimums of the various microbes involved in the process. Of course, it should be remembered that unless a closely controlled digester is used, a uniform temperature does not prevail throughout the mass of composting material at any one time – except at the start and end of the process when all material is at ambient temperature.

At temperatures lower than 30°C, a straight-line relationship exists in terms of increase in the efficiency and speed of the process and increase in temperature. The rate begins to taper off when the temperature passes 30°C and begins to approach 35°C. The slope of a curve showing efficiency or speed of the process as a function of temperature would be practically a plateau between 35°C and about 55°C – perhaps with some declination between 50 and 55°C. The existence of a plateau from the mesophilic range in terms of rate of activity is due not only to the involvement of many types of organisms but also to adaptation of organisms or enrichment for organisms adapted to a given range. As the temperature exceeds 55°C, efficiency and speed begin to drop abruptly and become negligible at temperatures higher than 70°C. At temperatures above 65°C, spore formers rapidly begin to assume the spore form and, as such, show little activity. Most of the nonspore formers simply die off.

It generally is said that as far as bacterial activity is concerned, a zone of repressed activity exists between the end of the mesophilic range and the beginning of the thermophilic range. The reason for this conviction is a supposedly sharp division of organisms into mesophiles and thermophiles. With respect to facultative organisms, it is thought that the alternative range is within either the strictly mesophilic or the strictly thermophilic ranges. However, results of studies by M.B. Allen[13] on the temperature requirements of aerobic spore formers and by this author[14] on the effect on anaerobic digestion showed that the zone does not occur in practical operations. The reason is that in such operations applied conditions are not as closely controlled and recording techniques are not as sensitive as is the case with small-scale laboratory studies. Furthermore, in laboratory experiments, a limited and controlled variety of microorganisms generally takes part in an activity at any one time.

There is some debate as to the relative merits of thermophilic vs mesophilic composting with respect to nature, extent, and rate of decomposition.[15] The argument is not peculiar to composting but is applicable to other biological waste-treatment processes as well. It is often assumed that in effect thermophilic organisms are more efficient than mesophiles. The rationale for the assumption is largely based on the intuitive belief backed by some experimental evidence that metabolic activity becomes more efficient at higher temperatures than at lower because they ultimately are the result of chemical (i.e., biochemical) reactions "catalyzed" by enzymes. Experimental evidence shows that up to a certain point chemical and enzymatical reactions are accelerated by each increment in temperature. For enzymes, the maximum is that point above which they begin to be inactivated.

There is reason to believe that the upper level of the optimum range of the thermophiles involved in composting is between 55 and 60°C. Undoubtedly, the process becomes less efficient when the temperature exceeds 60°C if for no other reason than that spore-forming organisms begin to enter the spore or resistant stage. When microbes are in the spore stage, activity diminishes practically to zero, and hence the composting process is correspondingly slowed. Another piece of evidence in favor of a drop in activity above 55 to 60°C is the limitation of the depth of the layer of actinomycetes and fungi to a zone in which the temperature is lower than 55°C. (Of course, this evidence is qualified somewhat by the fact that a lack of oxygen may be a contributing factor to the limitation.) Inasmuch as fungi and actinomycetes play an important part in the decomposition of cellulose and resistant material, decomposition temperatures inhibitory to them *ipso facto* result in a lowering of the efficiency of the compost process.

In a practical situation involving windrow composting, the question remains moot inasmuch as stated previously the temperature of a reasonably large or insulated mass inevitably reaches thermophilic levels unless something is drastically amiss with the operation. Therefore, the question has a significance mainly for mechanized (closed) operations, since with them heat is dissipated by the continued agitation of the material in the compost unit. Of course, temperature can rise in the units if the mass of composting material is sufficiently large and the degree of agitation of the contents is kept below a critical level.

The discussion on temperature should be concluded by pointing out that in a practical operation the desired temperature range should include thermophilic temperatures. The reasons are that (1) some of the organisms involved in the process have their optimum temperature in the thermophilic range; (2) weed seeds and most microbes of pathogenic significance cannot survive exposure to thermophilic temperatures; and (3) unless definite measures are taken, a composting mass of any appreciable volume will assume high temperatures.

- *Hydrogen Ion (pH level)* — The rationale proposed for the broad permissible temperature range also applies to the pH level. Generally speaking, one can state that fungi tolerate a wider pH range than do bacteria. The optimum pH range for most bacteria is between 6.0 and 7.5, whereas for fungi it can be between 5.5 and 8.0. In fact, the upper pH limit for many fungi has been found to be a function of precipitation of essential nutrients from the growth medium, rather than any inhibition due to pH *per se*.

In a practical operation, little can be done, or rather should be done, to adjust the prevailing pH level of a pile. Generally, the pH begins to drop at the initiation of the compost process. This is a consequence of the activity of the acid-forming bacteria which break down complex carbonaceous material (polysaccharides and cellulose) to organic acid intermediates. A part of the acid formation may take place in localized anaerobic zones. A part may be due to conditions of abundance of carbonaceous substrate and the resulting accumulation of intermediates formed by shunt metabolism arising from the abundance and perhaps from interfering environmental conditions. Regardless of the causes, the pH level may drop to as low as 4.5 to 5.0 with municipal refuse and perhaps lower with other wastes. While microbial activity may be inhibited somewhat at these levels, the effect is only transitory. Fortunately, the synthesis of organic acid is accompanied by the development of a population of microorganisms capable of utilizing the acids as a substrate. The net effect is that after a few days in a rapid-type compost process the pH begins to rise. The rise continues until a level of 8.0 to 9.0 is reached, and the mass becomes alkaline in reaction. Whether as a result, or by coincidence, the actinomycetes and fungi become grossly visible during the time of the high pH level.

In addition to the very sound economic reason for not using any unnecessary additives, which always entails an added expense, an important one for not trying to adjust the pH during the acid-forming period is related to the promotion of nitrogen loss. In experiments conducted at the University of California on the use of lime to prevent the initial drop in pH, nitrogen loss always was greater from piles to which lime $(Ca(OH)_2)$ had been added to raise the pH.[16] (Calcium acts as a buffer as well as an easy means of adjusting the pH to a higher level.) The normal nitrogen loss is increased by the elevation of the pH level at a time earlier than that in the absence of lime.

An exception to the rule of not attempting to prevent the drop in pH would be made in an operation in which the raw material is rich in sugars or other readily decomposed carbohydrates. The aerobic (and anaerobic) breakdown of such materials is accompanied by acid formation more extensive than occurs with municipal refuse. In studies on the composting of fruit waste solids mixed with sawdust, rice hulls, or composted refuse, adjustment of pH by adding lime eliminated the three- to four-day lag in temperature rise characteristic of unbuffered material.[17] Again, however, nitrogen loss was greater.

● *Aeration* – Probably a designation more appropriate than aeration would be oxygen availability. Aeration simply is the supplying of oxygen to the microorganisms. The rationale for maintaining an aerobic system was mentioned in the section on classification. Since modern composting is aerobic, oxygen supply becomes an important environmental factor. Of course, the implication of the preceding sentence is that with aerobic composting comes the collection of benefits mentioned earlier, namely, lack of objectionable odors, rapid decomposition, high temperatures, etc. The extent to which these benefits are attained is a function of the completeness with which the oxygen requirements of the microbes are met.

Before discussing oxygen requirements, it might be well to describe methods of aerating composting material. Although this sequence of topics may not at first glance appear to be logical, in the present instance it is because method of aeration determines the amount of air that must be applied in a given operation to supply the requisite oxygen. In some mechanical systems, aeration is carried out by forcing air through the composting material. In others, aeration is accomplished by a combination of stirring and tumbling of the material. In windrow composting involving municipal refuse, initial aeration is a result of the milling and stacking of the piles. The stacking operation itself is the means of initial aeration with those wastes not requiring milling as a pretreatment. Subsequent aeration is accomplished by turning the windrowed material. The oxygen used by the microbes in the windrows comes mostly from the air entrapped in the interstices (voids) in the pile. A very small amount enters through the diffusion of ambient air into the outer layer of the windrows.

The precise amount of air to be introduced into the composting material has long intrigued many investigators. The determination of the amount is not amenable to the relatively straightforward analysis used to arrive at the oxygen demand of a given wastewater (e.g., B.O.D. and C.O.D.). Schulze[18,19] was one of the earlier researchers on composting to be concerned with this aspect. In his investigation, he followed the simple expedient of enclosing the composting material and then forcing air through the chamber at a given rate and measuring the oxygen content of the air exiting from the chamber. While this approach did not lead to a determination of the total oxygen needed to aerobically compost the material, it did provide a measure of rate of oxygen consumption. Schulze found that the respiratory quotient $\left(\dfrac{CO_2 \text{ produced}}{O_2 \text{ consumed}}\right)$ remained at one. In his studies he attempted to quantify the relation between oxygen uptake and the effect brought about by varying the key environmental factors: moisture content and temperature. Qualitatively, his results were as one would expect: the more favorable the environmental conditions, the greater the oxygen uptake both in rate and amount. Obviously, the converse is true with unfavorable conditions. For example, Schulze noted that oxygen uptake increased from 1 mg/gram volatile matter at 30°C to 5 mg/gram at 63°C.[19] Other workers arrived at different numbers, as is to be expected because of the variability of solid wastes. Chrometzka[20] states oxygen requirements ranging from 9 mm^3/gram/hour for ripe compost to 284 mm^3/gram/hour for fresh compost. "Fresh" compost (seven days old) required 176 mm^3/gram/hour. Lossin[21] reports average chemical oxygen demands ranging from almost 900 mg/gram on the first day of composting to about 325 mg/gram on the twenty-fourth day. Regan and Jeris,[22] in a review of the decomposition of cellulose and refuse, noted the oxygen uptake at various temperatures and moisture contents. The lowest uptake, 1.0 mg/O_2/gram volatile matter/hour, took place when the temperature of the mass was 30°C and the moisture content was 45 percent. The highest uptake, 13.6 mg/gram volatile matter/hour, occurred when the temperature was 45°C and the moisture content 56 percent.

The variety of values cited in the preceding paragraph emphasizes the difficulty in determining actual oxygen requirement rates. The difficulty stems from the influence exerted by temperature, moisture content, size of bacterial population, and availability of nutrients on oxygen uptake. The reason for the variation is that O_2 uptake (i.e., demand) reflects biological activity and, in fact, is a measure of it. If one wishes to translate oxygen uptake into amount of aeration required to assure an adequate supply of oxygen when it is needed, the complexity of the problem is magnified because, in addition to the preceding factors, the aeration equipment and physical nature of the composting material must be taken into consideration.

The upshot of the preceding problems is that while determining absolute oxygen requirements of a composting mass is an interesting research undertaking, the findings are strictly limited in their application — limited to the material being tested and to the conditions of the test runs.

The ultimate amount or total theoretical amount is determined by the amount of carbon to be oxidized. Therefore, in designing for air flow through a mechanical digester, the indicated procedure would be to estimate the carbon content of a given mass and from that arrive at a number for the oxygen requirement. Such a procedure would result in an over-design, because obviously not all of the available carbon is oxidized in a biological process. On the other hand, some over-design is necessary because it would be practically impossible to so efficiently aerate a mass of wastes as to ensure the simultaneous contact of all microorganisms with the precise amount of oxygen needed by them at a single instant in time. Perhaps the air-flow requirement estimated by Schulze (562 m^3 to 623.4 m^3/metric ton [18,000 to 21,000 cu. ft./ton] volatile matter/day[18]) could be of some use, excepting that these numbers are based upon the use of his particular equipment and on a laboratory scale experiment.

The upshot is that a certain amount of pre-experimentation is needed with each type of mechanical digester to determine the required air flow within the unit, i.e., efficiency of the aeration device. In operation, the aim should be to adjust the air flow such that at least a small amount of oxygen remains in the exit air stream. The aforesaid problems of determining oxygen requirements do not apply to windrow composting, inasmuch as oxygen is not supplied by forced aeration except in special applications.

A very gross and yet effective means of monitoring for adequacy of oxygen supply is by way of the olfactory sense, namely, detection of odors. The emanation of putrefactive odors from a composting pile or digester is a positive indication that conditions have become anaerobic in it. The intensity of the odors is an indication of the extent of anaerobiosis. The odors disappear following an increase in the air flow through a digester or a more frequent turning of a windrow. While detection of objectionable odors may seem a rather gross monitoring device, it is an effective one for most applications. It does have the disadvantage of being an "after-the-fact" indicator. Therefore, in operations in which it is possible to measure the oxygen in the input and output air-streams, a direct monitoring is advisable.

A fact to keep in mind, however, is that even though oxygen may be present in the discharge air stream, localized anaerobic zones may be present because of inadequate mixing or tumbling in a digester unit. In other words, the mere presence of sufficient oxygen in an air stream does not necessarily ensure that all bacteria have access to oxygen. To secure complete aeration would be an undertaking more difficult than that with the activated sludge process in sewage treatment, and probably more expensive. It would involve reducing all particles to a size less than a millimeter or two because by its very dimensions a particle any larger could be anaerobic in its interior. Fortunately, complete aeration is not necessary to ensure a nuisance-free operation. Nevertheless, it must be adequate, i.e., sufficient to assure the absence of objectionable odors. Any gain from complete aeration as compared to adequate aeration would be outweighed by the additional costs involved.

• *Moisture Content* — Moisture content is discussed directly after aeration because with windrow composting the two are interdependent. This is not the place to delve into the reasons why a certain amount of water is necessary for the activity or even the survival of microorganisms and, in fact, of all living beings. Therefore, moisture here is discussed in terms of minimum and maximum requirements and of the circumstances that affect the extent of those requirements. With composting and other biological reactions, theoretically the ideal moisture content would be one that approaches 100 percent and would be attained by slurrying the wastes, because then moisture content obviously would not be limiting. However, in a real or practical situation other factors enter that make a 100 percent moisture content unsuitable. Chief among those factors is oxygen availability. In fact, the relation between optimum moisture content and oxygen availability is very close in composting, so much so that a discussion of one must entail the discussion of the other.

As stated earlier, the provision of the necessary degree of oxygenation in a slurry would be economically unfeasible in composting. Consequently, since a 100 percent moisture content is impractical, the discussion of optimum moisture content becomes one on the maximum permissible moisture content — how high it can be without leading to anaerobiosis. The maximum permissible percent moisture is a function of the physical nature of the material to be composted and hence varies with the wastes to be processed. On the other hand, the lower end of the optimum moisture content spectrum is fairly constant regardless of the nature of the wastes because it relates directly to the physiological needs of the microbes and is independent of the influence of some other factor. Inasmuch as generally all biological activity ceases at a moisture content of about 12 percent, the closer the moisture content of a waste approaches this level, the more retarded will be the rate of microbiological activity. In practice, the moisture content of composting material should not be allowed to drop below 45 to 50 percent.

The relation between structure and maximum permissible moisture content is especially important in windrow composting. If one pauses to consider the "morphology" of a windrow, one will realize why this is so. When a pile is set up, unless it has been deliberately compacted, it will have a significant fraction of its volume in the form of voids, i.e., the interstices between the particles as shown in Figure 1. The size of the interstices is determined partly by that of the particles, partly by their configuration, and partly by the "structural" strength of the material. Large, firm, irregularly shaped particles lead to the formation of interstices of a greater volume than would be possible with small, smooth-surface, regularly shaped particles.

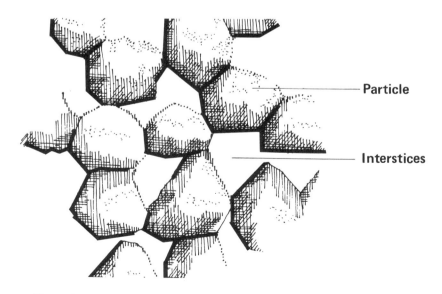

Particle

Interstices

Figure 1: Illustration of interstices in windrowed composting material

Interstitial volume is practically nonexistent with amorphous material. Certain substances such as paper quickly lose their rigidity when moistened and tend to collapse into amorphous, closely packed blobs. Thus, a pile composed largely of paper undergoes a loss in interstitial volume in proportion to its moisture content. For example, a pile consisting of 50 percent shredded paper and 50 percent vegetable trimmings (by volume) and with a moisture content of 70 percent will settle to less than a third of its original height within a 15- to 16-hour period. The loss in interstitial volume, and hence available oxygen, necessarily would be on a comparable level. On the other hand, a pile consisting to a significant extent of straw, sawdust, or rice hulls settles only slightly even at high moisture contents. In fact, the greater part of any settling would be due to weight of the mass causing a rearrangement of the particles. Since the oxygen used by the microorganisms comes from the air trapped in the interstices, the greater the portion of the interstitial volume occupied by water, the less that remains for "stored" air.

In view of all of these facts, the conclusion must be that the maximum permissible moisture content depends upon the extent to which the cumulative volume of the interstices can be filled with water and yet leave sufficient space to contain enough air to meet the oxygen needs of the microbes. Certainly, if the voids are completely filled with water, the pile becomes anaerobic. It is an elementary corollary that to remain aerobic, a pile that settles extensively can contain less water than one that settles hardly at all. Therefore, the maximum permissible moisture content in the former will be less than that in the latter. Examples of the variation that results are given in Table 2, in which are listed the maximum permissible moisture content of several wastes.

Another phenomenon related to moisture content is the effect of moisture on oxygen demand. In a report Lossin states that moisture content is a determinant of oxygen needs.[21] For example, fresh compost having a moisture content of 45 percent required 263 mm^3/gram/hour; whereas at a moisture content of 60 percent, the demand was 306 mm^3/gram/hour. This increase in demand indicates that an inadequate supply of moisture must have been a limiting factor for bacterial activity. It would have taken another determination at a higher moisture content to determine

TABLE 2. MAXIMUM PERMISSIBLE MOISTURE CONTENTS

Type of Waste	Moisture Content % of Total Weight
Theoretical	100
Straw[a]	75–85
Wood (sawdust, small chips)	75–90
Rice hulls[a]	75–85
"Wet" wastes (vegetable trimmings, lawn clippings, garbage, etc.)	50–55
Municipal refuse	55–65
Manures (without bedding)	55–65

[a]Serves as the "absorbent" and principal source of carbonaceous. Does need sufficient nitrogenous wastes (e.g., manure, grass clippings, sewage sludges) added to bring the C/N ratio down to the proper level.

whether moisture was no longer limiting at 60 percent. Incidentally, these observations indicate that with all other factors at optimum level, rate of oxygen uptake could be used as a means for determining optimum moisture content.

The moisture determination can easily be made. An aliquot of the raw material is weighed and then dried at 100°C for a period of six to eight hours. The dried material is weighed. The loss in weight over the original weight times 100 percent equals the percent moisture. Water can be added at any time in a mechanical digester. In windrow composting, it is better to add water during turning, inasmuch as the windrows tend to shed water when it is sprinkled on them. A crude indication of adequacy of moisture is a glistening appearance of the composting material. Excess moisture is indicated by the presence of drainage. Drainage may occur before foul odors become noticeable. A dry, dusty appearance indicates insufficient moisture. These latter indicators are qualitative in nature and should be backed by quantitative analyses. A drop in temperature is another indication of inadequate moisture, providing other factors are suitable.

THE PROCESS

Introduction

The steps in the compost process are those pertaining to the course of action pursued in the conduct of the compost operation — sorting, grinding, composting, and final processing. Of the four, sorting and grinding are preparatory steps. The sequence of the sorting and grinding steps depends upon the nature of the operation. Needless to state, the carrying out of the steps also varies with the nature of the operation and especially with the types of material to be composted. With most animal wastes, the preparatory steps need be only minimal and are concerned with moisture content and structural strength of the material. On the other hand, the preparatory steps are quite extensive for composting municipal refuse.

The present section deals with the rationale for the steps rather than with the technology involved in accomplishing them. The technology is reserved for a later section.

Sorting

Synonyms for *sorting* are *segregation* and *classification.* Sorting is, as the name implies, the removal of material not suited or not intended to be composted. In an operation involving agricultural wastes or other special homogeneous wastes, the sorting step is negligible or may even be omitted. On the other hand, the operation is a complex and expensive one when municipal wastes are involved. The remainder of this section on sorting concerns the step as it relates to composting municipal refuse.

As anyone who has any knowledge of municipal refuse is aware, one of the outstanding characteristics of the material is its heterogeneity. This is not surprising inasmuch as refuse is an accumulation or conglomeration of the discards of practically the entire spectrum of human activity. Consequently, it is a mixture of a wide variety of materials, some readily compostable and some completely resistant to biological attack and hence to composting. The mixture is not only one of a diversity of materials but also one in which the diverse materials are thoroughly and almost hopelessly intermixed.

An idea of the diversity, as well as of the relative amounts of the different residues, is indicated by the data in Table 3. Of the components listed in the table, the likely candidates for composting are garbage, garden debris, rags (natural fibers only), wood, and paper — all of which constitute the organic fraction of refuse. Although the remaining (nonbiodegradable) components would not interfere with the biological phase of the process, they could detract from the utility of the compost product and certainly would interfere with the mechanical aspects in terms of added volume to be handled and of wear-and-tear and clogging of the equipment. The only recourse, therefore, is to remove those components.

TABLE 3. COMPOSITION OF MUNICIPAL REFUSE

Refuse Category	% of Total Weight[a]
Paper	28–30
Paperboard	7–24
Garden wastes	7–35
Garbage	2–9
Metal	6–10
Aluminum	0.3–1
Glass	3–10
Cloth	1–3
Plastics	1.5–3
Fats and oils	2–6
Residue (ashes, dirt, etc.)	3–20

[a]Range of reported values

Another reason for removing noncompostable materials, and even some of the compostable fraction, is that of resource recovery. In fact, when resource recovery is involved, composting becomes but one activity in a complex of activities all geared to reclaiming residues to serve as resources. In such an operation, some or all of the paper may be removed as a part of a fiber-recovery operation either for paper manufacture or for conversion into heat energy. Two other reasons for removing a compostable component are to increase the efficiency of the compost phase and to improve the nutritional quality of the product for crop production. In both cases it is the carbonaceous fraction that usually is affected in that carbon is removed to bring down the C/N ratio of a waste to a more suitable level.

Noncompostable components may interfere with the operation by damaging equipment either through direct breakage or by wear-and-tear. An example of the former would be the breaking of a hammer in the hammermill by a piece of steel reinforcement rod accidentally slipped into the mill. An example of the wear-and-tear is the abrasion of hammers and grates caused by grit and possibly by dry paper and ground glass in the refuse stream. Adverse effects of nonbiodegradable materials such as glass and plastics are mostly by way of detracting from the quality of the compost product. The presence of glass shards, even of minute bits, in the compost product tends to limit use to field applications, although this is not necessarily a serious constraint. The home gardener would object to the presence of glass primarily for reasons of safety. Compost having glass bits in it should not be spread on pasture land because of the possibility of grazing livestock ingesting glass along with the grass. Plastic, especially film plastic, has the unfortunate tendency to accumulate on the surface of a field and thereby make the field unsightly. However, it has no adverse effect on crop yield. Nevertheless, as with glass, compost containing pieces of plastic should not be spread on pasture land. The plastic remains in the rumen of a grazing animal. Eventually, enough accumulates to result in serious and even fatal consequences to the animal.

Grinding (Size Reduction)

Synonyms for *grinding* are *size reduction, comminution, shredding,* and *milling.* They are used interchangeably in this text. If sorting is accomplished manually, then grinding follows. On the other hand, if sorting is done mechanically, grinding precedes sorting because of the fact that existing mechanical sorting systems are based upon a particle size much smaller than that of unmilled refuse. Although imparting an initial aeration and accomplishing some homogenization are important reasons for grinding refuse for composting, the primary purpose is to increase the surface area of the material, inasmuch as the smaller the particle, the greater is its ratio of surface area to mass. Therefore, the more finely a mass of material is milled, the greater will be the total surface area exposed to bacterial attack. It is axiomatic that the speed of a reaction is in direct proportion to the amount of surface exposed to the reactive agent or force, regardless of whether it be biological, chemical, or physical in nature. As far as biological reactions are concerned, the reason is that a greater surface area exposed to bacterial attack results in access to nutrients by more microbes per unit of time and hence accelerated decomposition. Of course, in the process of particle size reduction, natural barriers such as waxy or other resistant coatings and surface layers are ruptured, thereby permitting microbial ingress to the interior of the particle.

While theoretically it may be true that the smaller the particle size, the closer is the approach to optimum, upper and lower limits exist in practice beyond which the process begins to be unduly slowed or rapidly becomes economically unfavorable. Aside from the dictates of economics, optimum particle size in the average compost operation is a function of the structural strength of the raw material. The firmer (i.e., the more resistant to compaction and settling) the raw material is, the smaller can be the particle size without redounding to the detriment of the process. Thus, the optimum particle size for woody material, straw, wood shavings, and like substances is a matter of a centimeter or less. On the other hand, material such as paper, and especially garbage (food preparation wastes) and vegetable trimmings, should not be reduced to a particle size *less* than 2.5 to 5 cm (1 to 2 in.). The reason for the difference between the two types of materials with respect to minimum size ultimately is the relation of interstices between particles to aeration requirements. A structurally strong material retains its shape when wetted even though it had been reduced to a very small particle size. A structurally weak material would compact when stacked and collapse when moistened. Thus, with the former material the integrity of the interstices would not be affected; whereas with the latter, it would be lost.

With respect to maximum permissible particle size, a safe range for municipal refuse is 2.5 to 5.0 cm (1 to 2 in.). This size range is close to that of the output of most grinders presently on the market. Maximum particle size is less important when easily decomposed material is composted. For instance, this author produced a very high quality compost within 12 days with unmilled vegetable trimmings and fresh garden debris as substrates. The extent of size reduction consisted in breaking overly long flower stalks into lengths that would fit within the bin in which the composting was done.

Economics is an important factor in determining a practical particle size in composting. At each pass through a grinder, a gradation and corresponding ratio of particle sizes up to a certain maximum is encountered. The ratio of smaller particles to the larger ones can be increased by passing the milled refuse through the grinder a second time. To do so would not necessarily double the energy requirement, but it would materially increase it, providing the same size grate openings were used. The reason for the failure to double the power (energy) need is that along with certain other factors, energy requirement is a function of relation of particle size of the material *fed into* the grinder *to that of the output*. However, it should be remembered that if the same size grate openings are used in both passes, the particle size distribution (i.e., range of particle sizes) of the milled refuse will be the same with both passes. Nevertheless, the fraction of the output from the second pass in the smaller end of the size distribution pattern will be greater than that in the first pass. Repeated grinding will *not* lead to a uniformly sized product — there will always be a size distribution.

The effect of input particle size on energy consumption was shown in a study by Trezek and Savage[23,24] involving the use of a 9 metric ton/hour (10 ton/hour) Gruendler hammermill. They found that to comminute commercial wastes to a size distribution equivalent to that accomplished with residential refuse about 40 percent more energy would be required. The reason was that the characteristic particle size of the commercial refuse used in the study was about twice that of the residential refuse, namely, 15 cm vs 7.2 cm or 6 in. vs 3 in. (The characteristic particle size is the size corresponding to 63.2 percent of the cumulative particles passing through a screen of size x.) The difference in energy requirements was tantamount to passing the commercial refuse twice through the hammermill. The example cited by the two researchers involved an energy requirement of 58.05 Mj/metric ton (14.5 kwh./ton) with the residential refuse and 81.7 Mj/metric ton (20.5 kwh./ton) with the commercial refuse to reduce the respective wastes to a characteristic size of 2.35 cm (0.96 in.). The actual numbers would vary from operation to operation, inasmuch as they are a function of other factors in addition to that of particle size. The points to note from the preceding statements are: (1) The characteristic particle size distribution of the product with both materials remained the *same* because the grate opening size was not changed. (2) To produce a smaller characteristic size, the second pass must be accompanied by the installation of smaller grate openings in the hammermill. If the latter is done, then the energy cost would approach that of the primary pass, depending upon the similarity of the ratio of input to output particle sizes in both passes. For example, if the ratios of the input to output particle sizes in the two passes were 2:1, the energy requirements would be comparable for each pass. Energy requirements sharply increase with reduction in characteristic particle size. Trezek and Savage[23] report that with the characteristic particle size of the input refuse at 15 cm, an energy consumption of about 87.2 Mj/metric ton (22.1 kwh./ton) to produce a characteristic particle size of a little more than 0.25 cm (0.1 in.) to about 3.96 Mj/metric ton (1 kwh./ton) at 7.5 cm (3 in.). It should be noted that the relationship is not straight line. The rate of decrease in energy required to produce a given characteristic particle size tapers off as the particle size becomes larger, i.e., greater than 5.1 cm (2 in.) and approaches 15 cm (6 in.).

Aside from machine design and characteristic particle size, other factors that affect energy consumption are moisture content of the waste, feed rate, and rotational velocity of the hammers. Power requirements are least at a moisture content between 40 and 50 percent and greatest at 25 and 60 percent, the highest moisture content tried.[24] Energy consumption is less at lower rotational velocities of the hammers, but the savings are offset by a coarser output (larger characteristic size).

Composting Step

In this section the interest is in the phenomena that occur during the compost step rather than in the technology to accomplish it. The composting process, whether it occurs in an enclosed unit (digester) or in an open windrow, is characterized by certain phenomena and follows a definite course. Allusion to some of these aspects of the process has been made in the preceding section. The principal features are: (1) a rise and fall in temperature, (2) uptake of oxygen, (3) a change in appearance and odor of the composting material, (4) a lowering of the C/N ratio, (5) an initial drop followed by a rise in pH level, and (6) the production of ammonia. Of these features, change in pH level and production of ammonia were discussed in the section on environmental factors and thus are given no further attention here.

■ **Temperature Rise and Fall:** The rise and fall in temperature is one of the outstanding characteristics of the compost process and is much more apparent in windrow composting than in mechanical composting; but even in the latter, it can be detected by a lessening of the operation's heat input requirements. In fact, the characteristic temperature pattern is one of the more important features in monitoring an operation, since any deviation from the characteristic pattern betokens an inhibition of the process. The ultimate reason is that the rise in temperature is due to bacterial activity. Therefore, any decline in bacterial activity is accompanied by a drop in temperature.

The temperature begins to rise directly after the material is ground and stacked in a windrow or placed in a digester. Generally, there is a very short lag period — often so short as to escape detection. If no lag period occurs, it is due to the fact that decomposition has already begun at the time the material was discarded and thus became a waste. The more readily decomposable a material is, the more advanced will have been the degree of its decomposition by the time it is processed for composting. Except for brief interruptions during turning in the windrow method, the rise in temperature continues unabated until the 50 to 55°C level is attained. At this point, the rate of ascent begins to taper off until a plateau is reached at 60°C. Thereafter it hovers between 60 and 65°C and occasionally may peak at 70°C and rarely higher. Such high temperatures may not be reached in a mechanized operation because the continuous tumbling of the material results in a continual loss of heat. After some time, the temperature begins to decline regardless of how close to optimum operational conditions are maintained. Eventually the temperature drops to that of the ambient temperature. A typical temperature curve for a "high-rate" system (University of California method) is presented in Figure 2. With slower systems the rate of increase may be less precipitous, but the configuration of the curve is that of the figure.

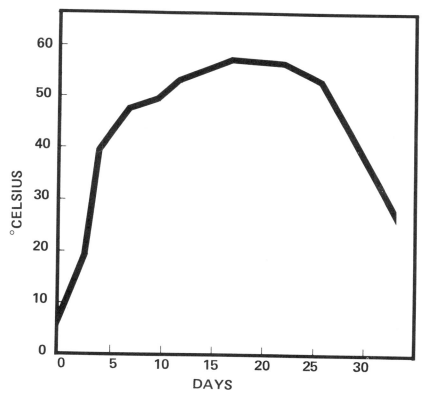

Figure 2: Temperature curve for a windrow of composting material

The initial rapid straight-line rise in temperature is a result of the high rate of decomposition made possible by the availability of easily decomposable material in the waste. This of course assumes that other environmental conditions, especially aeration and moisture, are at optimum levels. As the readily broken down substances (starches, sugars, proteins, amino acids, etc.) are used up and only the more complex compounds remain, the intensity of bacterial activity wanes, heat is generated at a less rapid rate, and consequently the rate of temperature increase of the composting mass tapers off, accounting for the plateau at the middle of the curve. Eventually only the more resistant compounds remain, and the temperature begins a final and inevitable drop.

The two principal factors that can lead to a deviation from the typical temperature pattern are an inadequate supply of oxygen or insufficient moisture. If the temperature deviation is due to insufficient oxygen, that fact will be indicated by the emanation of foul odors from the composting material. Lack of oxygen in the discharge air is the indicator in a mechanized digester. When the problem has been remedied, the temperature will resume its normal course. If inadequate moisture is the cause, it can be detected visually or, of course, by laboratory analysis. The addition of water brings the temperature back to its normal level. It should be noted that excessive moisture is not designated as a cause of temperature drop. Rather, it is an indirect cause in that because of it oxygen is excluded; hence, strictly speaking, insufficient oxygen becomes the direct cause.

The temperature is not at a uniform level throughout a windrow. Characteristically, it rises with increase in distance from the exposed surface area of a pile. Thus, the temperature is highest in the center zone, unless this zone happens to be partially or completely anaerobic. Naturally, there is no sharp demarcation between zones, and instead one blends into another. Nor are the zones uniform in thickness. For example, in a large windrow pockets may be found in which the temperature level may be higher or lower than that in the remainder of a given layer.

The degree of proximity of the high temperature zone to the surface of a pile is a function of ambient temperature, "porosity" (and therefore moisture content of the outer layer), and availability of oxygen. Unless the piles are completely insulated, the ambient temperature will have some influence on the temperature of the surface layer of the pile. Nevertheless, because of the insulating character of refuse, high temperatures are reached in the interior of piles that may be exposed to ambient sub-zero temperatures. The porosity of the outer layer determines the extent and depth to which ambient air penetrates a pile as a function of wind velocity. Hence wind velocity also is a factor. Moisture content exerts its influence in two ways: First, by filling the interstices, it decreases the amount of ambient air that can penetrate the pile; second, it brings about an increase in convection of heat from the high temperature zone. During peak activity, temperatures of 50°C may be encountered to within 5 to 30 cm from the surface layer, depending upon the factors mentioned. The high temperatures in the center of the pile may extend down to its base.

Any suitable temperature-measuring device may be used to ascertain the temperature. For monitoring purposes, the temperature should be determined at several points in a windrow and at two or three depths. Suggested depths are 10 cm (4 in.), 30.5 cm (12 in.), and 91 cm (36 in.). Most mechanized units are equipped with systems for measuring and recording the temperature.

■ **Oxygen Uptake:** This aspect was covered in the section on aeration and hence requires no further mention at this point other than to point out that increase and decrease in oxygen uptake parallels the rise and fall in temperature, and for the same reasons. As the bacterial populations become numerous and active, their oxygen need and consequent uptake become greater. Conversely, a decline in activity is accompanied by a drop in oxygen requirement.

■ **Change in Color and Odor:** Common to the composting of practically every type of material is a darkening in color as the process advances. Taking municipal refuse as an example, when freshly ground it is grayish green. After three or four days of application of a high-rate composting method, it becomes darker; and at the end of a week it is grayish black. Moreover, the identity of "green" garbage and garden debris is no longer apparent. However, paper and newsprint continue to be recognizable. When the actinomycete-fungal layer makes its appearance, paper other than newsprint begins to lose its identity. Newsprint retains its identity even after the active phase of composting has been completed. Straw, sawdust, wood chips, rice hulls, and other like materials are recognizable throughout the process, although their texture undergoes a considerable change. Straw loses its toughness and is readily broken. Wood chips and sawdust are darkened, partly through the absorption of the heavily pigmented humic acids, and their structural strength is somewhat lessened. The same fate befalls rice hulls. The sequence of color and structural changes remains the same regardless of compost method. The only difference is in the length of time required for the changes to take place.

The characteristic odor of the raw material being composted begins to disappear after a few days, the number being a function of the compost method. The type of odors in the succession of changes depends upon the raw material. Manure passes from its characteristic rank odor, through an ammonia phase, and eventually ends up having a not unpleasing earthy smell. With municipal refuse and garden debris, a period of aromatic ("cooking") smells intervenes between those of the original "garbagey" and the final earthy odors. Often, the onset of the earthy smell stage is immediately preceded by one in which the odor of ammonia is more or less pronounced.

■ **Change in C/N Ratio:** As the compost process progresses, the C/N ratio of the mass begins to decline. The decline is due to the formation and loss of CO_2. A portion of the biodegradable carbon in the composting material is assimilated by the microbes and converted to microbial protoplasm; while the remainder is oxidized to CO_2 by the microorganisms to meet their physiological energy needs. The CO_2 diffuses into the surrounding air, and thus the carbon content of the composting mass is lowered. Although some of the original nitrogen content may be lost during the compost operation, the amount does not match that of the carbon loss; therefore, the ratio of carbon to nitrogen declines. Since sawdust and other wood wastes have a very high carbon and low nitrogen content (and hence high C/N ratio) and at the same time break down very slowly, a composted product having a large wood fraction may have a higher C/N ratio than would be advisable with more readily compostable material, such as dry leaves, tissue paper, and wrapping paper. The only niche for C/N in monitoring is the determination of when the compost is ready for the soil.

■ **Determination of Degree of Stabilization:** By *stabilization* is meant the state or condition in which the composted material can be stored without giving rise to a nuisance or can be applied to the soil without causing problems there. This state is reached when the readily decomposed constituents have been oxidized to relatively stable intermediates. The term *stable* is qualified by the adverb *relatively* because stability in the strict sense would be an oxidation of the organic matter to CO_2, H_2O, and inorganic ash (the chemical elements making up the original material). As will become apparent in the section on uses of the compost product, this "ultimate" stability is not desirable because the utility of compost depends upon further decomposition in the soil. Rather, the desired degree of stability is one in which the readily decomposed compounds are broken down, and only the decomposition of the more resistant biologically decomposable compounds remains to be accomplished.

If the proper degree of stabilization is not attained, during storage the remaining putrescible material will continue to decompose, but not under controlled conditions and usually under anaerobic conditions with their concomitant production of foul odors. The problems that could arise from incompletely composted material in the soil are primarily related to crop damage. The damage may come from too high a C/N ratio or from the production of ammonia in the soil. Allusion to the effect of an excessively high C/N ratio already has been made and is discussed in some detail in the section on the use of the product. Ammonia above trace amounts is toxic to plant roots. Insufficiently composted material can be applied to the soil without damage to plants provided due precautions are taken; but to be aware of the need to apply these precautions, it is necessary to know whether the product has been sufficiently composted.

While the importance of determining degree of stability is widely recognized, the precise point at which stabilization is sufficiently far advanced is as yet to be agreed upon. The reason for the lack of a consensus is easily apparent. Because the value of the compost product depends upon additional decomposition in the soil, a range rather than a precise point is open to selection. For example, one group may hold that the material is sufficiently stabilized when the period of intensive handling (i.e., detention in the digester) has been completed. Entrepreneurs for mechanical systems usually advocate this criterion, the motivation being that they can thereby attribute a faster rate of composting to their equipment than could be done were a higher degree of stabilization required. Others more justifiably hold that the proper degree of stabilization is one at which the material can be stored without subsequent excessive heating or generation of bad odors. To justify their position, the backers of the shorter end of the stabilization scale divide the compost process into two parts: composting and maturation. Accordingly, *composting* would apply to the part of the process which takes place in the digester and *maturation* to that which occurs outside the digester. Theoretically, all of the more highly putrescible material is broken down in the composting stage, leaving the more resistant material for stabilization in the maturation step.

The proponents for the dual nomenclature call for some further processing — usually turning — during the maturation phase. Others regard such a dual nomenclature as being superfluous and potentially misleading. They hold that the term *composting* applies to the entire operation, i.e., from the time the material is milled and subsequently either placed in a digester or stacked in a windrow to the time at which it has become properly stable. *Stable* here is taken in the sense that the product can be stored without further treatment, yet not give rise to a nuisance, or can be applied on the land without causing crop damage.

Before a standard for stability can be established, the division of opinion described in the preceding paragraph must be resolved because standardization implies uniformity and ability to replicate.

Two important reasons come to mind for having a standard for evaluating degree of stability. One is to provide a common measure for comparing the efficiency of different composting systems. This is especially important in weighing the validity of the claims of a promoter for a particular piece of equipment or for his process. Allusion to this aspect was made in the preceding paragraph. The second reason is the matter of economics. The effect on economics comes in three ways: equipment costs, operational costs, and land costs. The shorter the time the material must be processed, the smaller the requirement for total equipment capacity; therefore the investment for equipment is correspondingly less. Similarly, less handling is required if the indicated composting time is shortened. Land requirements also are reduced, because less area is required for facilities and for windrows.

The problem of a lack of a suitable methodology has been an important contributing factor to the absence of a standard. However, several methodologies have been proposed. Before describing the methods, mention should be made of criteria that are not suitable. One is change of color and appearance in general. A composting material may be dark brown or grayish black and have the general appearance of a sufficiently composted product, yet it may have an excessively high C/N ratio or give rise to some very bad odors if stockpiled. The same strictures apply to using the possession of an earthy odor or present lack of odor as a sign of completed process. The earthy odor is characteristic of actinomycetes, and consequently is to be expected when those organisms are active. Since the actinomycetes require a decomposable substrate, their presence, at least at their first manifestation, would indicate the existence of unstable "organic" matter.

• *Methodology* — Among the several methods that have been proposed for determining the degree of stability are (1) final drop in temperature,[16] (2) degree of self-heating capacity,[25] (3) amount of decomposable and resistant organic matter in the material,[26] (4) rise in the redox potential,[27] (5) oxygen uptake,[19] (6) growth of the fungus *Chaetomium gracilis*,[28] and (7) the starch test.[29]

Use of the final drop in temperature to determine stability is based upon the fact that once the readily decomposable material has been decomposed, the high temperature prevalent during the peak composting period begins an inevitable descent. This phenomenon was described in the discussion on temperature in the sections headed "Principles" and "The Process." The finality of the decline can be demonstrated by the failure to bring about a resurgence in the temperature despite the application of optimum conditions. Moreover, it has been observed that once the temperature has dropped to 40 to 45°C, the material can be stored indefinitely without causing problems. Since this combination — drop in temperature with no-nuisance production — occurs at the attainment of a degree of stabilization compatible with a practical compost operation, temperature drop is a useful parameter for stability. It has the advantage of universality of application since the course of the temperature rise and fall (i.e., shape of the temperature curve) is the same regardless of the nature of the material being composted.

The self-heating capacity analysis proposed by Niese[25] may be regarded as a variation of the "final drop in temperature" parameter. According to Niese's method, representative samples of the material to be tested are placed in Dewar flasks. The flasks are covered with several layers of cotton wadding held in place by cellulose tape. A suitable sensing device is used to determine the temperature. To avoid loss of heat from the flasks, they are placed in an incubator in which the temperature is regulated by a temperature-difference sensing device. Degree of stability of the material is indicated by the extent of the rise in temperature. Niese's results indicate a temperature above 70°C with raw refuse, 40 to 60°C with material partly stable, and lower than 30°C with completely stabilized material. Niese's method has the advantage of the "universality" of application characteristic of the "final drop in temperature." However, it is a slow method and may require several days to complete. An advantage is that it does not depend upon all composting conditions being optimum at the time the determination is made. The "final drop" in temperature does involve some degree of uncertainty because it depends upon all conditions being suitable, a situation about which one can not always be assured.

Rolle' and Orsanic's method[26] was designed primarily to ascertain the amount of decomposable organic matter in wastes. Since it is used to measure decomposable material, it can be used to determine that of composting material. The difference between the numbers obtained with raw wastes and with composting wastes should indicate the degree of stability of the composting material. The principle upon which the method is based is the determination of the amount of oxidizing reagent used in the test. This is a measure of stability since the compost process is largely based on oxidation. The test consists of treating the sample with potassium dichromate solution in the presence of sulfuric acid. A certain amount of dichromate added in excess is thereby used up in the eradication of organic matter. The oxidizing agent left at the end of the reaction is back-titrated with ferrous ammonium sulfate, and the amount of dichromate used up is determined. The calculation of the amount of decomposable organic matter is as follows:

$$DOM = m^{\ell}N \left(1 - \frac{T}{S} \cdot 1.34\right)$$

in which DOM is decomposable organic matter, m^{ℓ} is ml of $K_2Cr_2O_7$ solution. N is normality of $K_2Cr_2O_7$. Quantitatively, resistant organic matter is that amount of loss by combustion not degraded in the oxidation reaction.

Möller[27] developed his oxidation-reduction potential method as a means for evaluating the "hygienic" condition of the compost product. His basis for such an evaluation is the reported finding of human pathogens and parasites in insufficiently composted material and the absence of such organisms in "mature" compost. Decomposable materials lead to an intensification of microbial activity and hence oxygen uptake and to a fall in oxidation-reduction potential. The oxidation-reduction potential rises as the organic material becomes mineralized. As a result of his work Möller states that the material may be regarded as being thoroughly fermented when the oxidation-reduction potential of the core zone of a pile is <50 mV lower than that of its outer layer. No information is given as to the relationship in a mechanical digester, in which no such zonation can occur.

An important handicap in the use of the redox-potential is that the measurement is not sufficiently accurate and is subject to a variety of interfering factors.

The measurement of growth and formation of fruiting bodies was used by Obrist and Allenspach[28,30] as a means of estimating existence of full maturity (stabilization) of composting material. Obrist states the special utility of the test rests upon the fact that the values obtained do not depend upon the components of the waste (*sic* household refuse). The growth is dependent upon the chemical nature of the wastes as a whole. In conducting the test, the fungus is cultured upon a solid nutrient medium containing a pulverized sample of the compost. After 12 days of incubation at 37°C, the fruiting bodies are counted. The substrate mycelium develops only on samples that have not reached an advanced stage of decomposition. The more advanced the degree of stabilization, the fewer will be the fruiting bodies. In his observations Obrist noted that a fruiting body count of less than 300 per 25 ml indicates a maturity sufficiently advanced to permit safe application on the field. A disadvantage in the use of the test is that it is time-consuming. Moreover, it is dependent upon the skill of the analyst; and according to Obrist, more observations are needed to standardize the test.

Chrometzka[20] and Allenspach[30] noted a relationship between oxygen uptake and maturity of the compost. Chrometzka used the Warburg apparatus to measure the direct oxygen requirement in mm^3/hour for 1-gram samples of the material to be tested. His studies indicated that under optimum conditions, the oxygen demand by fresh compost is 30 times that by mature compost. He cites Teensma as finding that the oxygen requirement for fresh refuse is 12 times that for well-fermented compost. The difficulty with using oxygen uptake as a measure of maturity is that it is relative, i.e., related to the composition of the material being tested. Hence, no absolute value can be given whereby a given compost can be termed stable if its oxygen demand equals that value.

Lossin's[29] starch determination method for estimating degree of stability rests upon the assumption that the starch content declines with increase in stabilization. Lossin states that he has always found starch in measurable quantities in municipal refuse. He speaks of three types of carbohydrates to be found in wastes, namely, sugars, starch, and cellulose. In windrow composting, sugars disappear within a week; and starch is used up by the fourth or fifth week of composting. The reasoning is that since starch is easily broken down and metabolized, and all wastes contain starch, it must be used up before the composting waste can be regarded as a microbiologically stable product. Lossin's test is based upon the formation of the starch-iodine complex in an acidic extract of the compost material. The test requires a considerable amount of care in conducting, especially in standardizing. Another problem, also characteristic of certain other tests discussed herein, is that of selecting truly representative samples. This is particularly difficult with municipal refuse because of its extreme heterogeneity.

Final Processing

When the composting step has been completed, the next step is that of final processing to prepare the product for marketing or for storage. If the material is to be stored, the final processing may take place before or after storage. Waiting until the end of the storage period permits further "curing" or maturing – a factor which in many cases redounds to the improvement of the product. The final processing may consist of regrinding and/or screening the material to increase the uniformity of particle size and eliminate oversize particles. In the Mexican operations, the use of the grinder did little to improve the physical characteristics of the product, whereas screening resulted in the desired uniformity and increased the "eye appeal" of the compost.

The use of a conventional shaker screen may be attended by problems, especially if the moisture content is high enough to cause the material to compact. In the Mexican operations blinding of the screens is lessened by blowing air through the bottom of the screen and thus somewhat "fluidizing" the material. In the University of California studies, "blinding" is minimized through the use of a trommel screen.

Unfortunately, the final screening does not remove the tiny bits of glass and plastic found in milled municipal refuse. The glass bits pose more a visual problem than a handicap in crop production. The pieces usually are minute and have rounded edges due to abrasion during milling and composting. The glass would be a problem mainly if spread upon pasture land. The plastic pieces detract from the appearance of the product, although they would exert no inhibitory influence on crop production. On the other hand, as stated before, if spread on pasture land and ingested with grass by grazing animals, the glass and plastic could have a dire effect upon the health of the animals.

Volume Reduction

At this point it would be well to discuss the extent of volume and weight reduction accomplished in the compost process. The extent of both is a function of the amount of readily decomposable material in the raw refuse and of the time allotted to the process. Obviously, the larger the rapidly decomposable fraction, the greater will be the extent of reduction. In his experiments this writer obtained a 60 to 70 percent reduction in volume when vegetable trimmings and garden debris served as substrates. The average reduction in volume and weight with municipal refuse was on the order of 30 to 40 percent. Some reports speak of a 46 percent reduction in weight of refuse. The reduction in volume and weight results from the conversion of some of the carbon to CO_2, which is then lost to the atmosphere. Some weight reduction may stem from the loss of moisture through evaporation. Loss in volume has a significance when composting is practiced solely as a waste "disposal" device. If the product is to be used for land reclamation or crop production, then the extent of reduction is important only in terms of amount available for use.

TECHNOLOGY

Introductory Remarks

Despite the perennial announcement of breakthrough innovations in the technology of composting, the major economically feasible advances have all been made. The so-called breakthroughs are either a minor modification of existing systems or the utilization of mechanical devices unrealistically expensive either in capital or in operating costs, or in both. The reader may have noticed this repeated reference to costs. The references are needed to stress the point that composting, especially in the United States, is by no means a lucrative operation. Consequently, economics places a very severe constraint on the elaborateness with which a practical composting operation can be designed. This constraint is especially onerous and frustrating to the chemical engineer and industrial biologist who can see a number of ways of improving existing systems. The trouble is that the trade-off between time-savings and the resulting reduction in equipment and handling needs and the cost of accomplishing them unfortunately is heavily weighted in favor of increasing the costs. However, it is fortunate that the potential gains from further sophistication of design are not exceedingly great. Where the real need exists is in lowering the costs of even the "low-cost" systems.

Size-Reduction Equipment

The design of size-reduction equipment rests upon three basic mechanisms, namely, impaction, tearing, and shearing. The emphasis on any one of the three depends upon the type of machine. Common to all existing designs is a strong dependence upon brute force to accomplish the size reduction, probably because originally none were designed to process municipal refuse: Most were intended to pulverize brittle materials. Refuse, on the other hand, is a conglomeration of materials ranging from soft to brittle. The result is that the existing size-reduction machinery is of necessity over-designed. Over-design brings with it the penalties of unnecessarily high capital and operating costs.

There are two principal types of grinders for municipal wastes: the hammermill and the ring grinder. The hammermill is the older of the two, and of the two it is the more dependent upon brute force. Generally, the term *hammermill* is applied to a machine in which flailing hammers impact material as it falls through the machine or as it rests on a stationary metal surface. The material is hurled against fixed surfaces or between the moving hammers and the outlet grate bar system. Some hammermills have a horizontal rotor shaft for the hammers, others a vertical shaft. With the horizontal type, size reduction begins when the hammers strike the material as it enters the rotor and when the material is pinched between the hammers and grate bars and is discharged from the machine. Oversize particles remain in the machine and are buffeted until they are reduced to a size sufficiently small to pass through the grate bars. In horizontal mills, maximum particle size is basically determined by the distance between the ends of the hammers and the housing and by the openings between the grate bars. The Gruendler hammermill is a well-known representative of this type of mill.

In a vertical hammermill, the rotor is positioned in a funnel-shaped housing and the material is impacted by the hammers during the fall to the bottom of the machine. Further reduction in particle size results from the pinching and crushing action brought about by the material falling through the increasingly smaller space between the edge of the hammers and the grinder housing toward the bottom of the machine. No grate bars are at the bottom, and the material simply drops out of the machine. (Some grinders may have a perforated housing around the rotor. Size reduction is accomplished by forcing the material through the openings. The material is then discharged from the bottom of the machine.) A well-known example of a vertical hammermill is the Tollemache.

The ring grinder is characterized by a vertical shaft placed in a funnel-shaped housing. The hammers are replaced by ring grinders held in place with pin-to-grinder clearance of 0.5 to 4 in. Because of this clearance, the ring grinders move outward as centrifugal force is applied. Size reduction is accomplished by attrition grinding. The best-known machine of this type is the Eidal grinder. The Eidal machine is equipped with rotating breaker bars at the top of the rotor. These perform an initial reduction through a tearing action, as well as prevent oversize material from reaching the ring grinders. The major reduction occurs as the material works its way downward between the rotating ring grinders and fixed bars projecting inward from the liner that houses the grinders. The material is ejected tangentially from the bottom of the machine. Maximum particle size is regulated by adjusting the distance between the "choke" ring and the ring grinders.

The advantages and disadvantages of each configuration of rotor, i.e., horizontal and vertical, have been thoroughly discussed in a three-part series of articles by Franconeri.[31-33] An important advantage of the horizontal configuration is the control over maximum output-particle size. This control is especially important in resource recovery systems — which includes composting and other biological systems for processing wastes. A primary disadvantage is that it cannot be freeloaded. The result is a considerable wear rate on the hammers, as well as the possibility of overloading or jamming, inasmuch as oversize or resistant material cannot be rejected by the machine. The flow rate of incoming wastes parallel to the rotor shaft allows for freeloading in vertical mills, and hence minimizes the difficulties resulting from the lack of it with horizontal mills. Overloading and jamming are eliminated in vertical mills by a combination of a funnel-shaped housing (broadest opening at the top) and special hammers at the top (prebreaking chamber), which results in the ejection of oversize and unmanageable items from the top of the machine. Other advantages are a lower wear rate, less chance of damage to the machine, and a lower housing and foundation requirement than is the case with horizontal mills. Disadvantages are (1) heavy wear and tear on the lower thrust bearing which supports the rotor; (2) difficulty of access to the rotor and other internal parts of the machine for maintenance and repairs; (3) difficulty in removing the rotor for necessary maintenance; and (4) very importantly, less control of the maximum particle size of the output. Particle size with a vertical mill is a function of composition of the input material and feed rate.

In the 1950s, grinders with basic designs different from the types described in the preceding paragraphs were developed in Europe. Of these, two were used fairly widely until the mid-1960s. One was the rasping machine developed by VAM in the Netherlands, and the second was the Dano "Egseter."[34] The rasping machine has a horizontal perforated bottom through which refuse is rasped by means of slowly rotating arms. The bottom is formed of alternating rasp plates and sieve plates. The latter have openings of a size to result in a milled product having a maximum particle size of 1 in. The wastes are moved over the plate by rotating arms. Small particles fall through the sieve plate. The rasp plates are made by welding manganese-steel square bars to a plate in a staggered pattern with bars spaced about 2 in. (5 cm) apart and protruding about 1.5 in. (3.8 cm) above the plate. The "Egseter" is a drum-type cylinder which rotates about 12 rpm about the axis of the cylinder. About 30 in. (75 cm) inside the outer shell are placed rough, hardened-steel bars. A screen or sieve is inserted between the rough bars and the outer shell. Size reduction is accomplished by the rasping action brought about by the pieces of wastes rubbing against each other and against the rough bars as the drum rotates. When the proper size is reached, the particles drop between the bars and through the screen.

The power requirements for both machines are low. The Egseter uses about 6 kwh./ton refuse. Since both machines are relatively simple in construction and involve no high rotational velocities, their construction is much simpler and maintenance problems much less severe than with the hammermill or versions thereof. The drawback is that neither machine can satisfactorily size reduce refuse that has an appreciable paper content, as is the case with the wastes generated in most cities in industrialized or developed nations. On the other hand, the two types are well suited for operations that involve wastes having an appreciable organic fraction and a minimum of paper, textiles, and plastic film.

Sorting

Sorting, whether manual or automated, applies principally to municipal wastes.

■ **Manual:** The simplest, but not necessarily the cheapest, method of sorting is manual sorting. Although with minor exceptions the high price of labor rules out manual sorting in the U.S., it remains the least expensive method in those nations in which labor is abundant and wages are low. For example, at the time of this writing, manual sorting was economically feasible in resource-recovery plants in Mexico.

The reliability of manual sorting obviously is dependent upon the workers. Potentially, it can be more reliable and effective than automated sorting because the human eye, acting as the sensor for the brain, brings to bear a complex capable of sensing and reacting to inputs at a rate and in a manner beyond the capacity of the most sophisticated sorting devices in existence today. The crucial factor in reliability is the worker, i.e., his cooperation. Cooperation can be brought about by training, motivation, and supervision. Training tells the worker what is expected of him or her. The importance of motivating is so obvious as to require no further mention. Supervision compensates for lapses in motivation and training.

Typically, the arrangement for manual sorting is as follows: The waste to be sorted is transferred to a conveyor belt which moves it past a magnetic separator to remove ferrous objects. Then it is transferred to a horizontal belt at which the "pickers" are stationed. In the Mexican operations, each worker has the responsibility of removing one or two types of objects (e.g., paper and glass) and of dropping the objects into nearby chutes that lead to containers placed on a lower floor. Each of the plants has two parallel belts for sorting. The sequence differs from the usual in that the intact cans and large ferrous objects are removed by hand and only milled material is exposed to the magnet.

■ **Automatic Sorting**: A sorting machine must be reliable, durable, easy to maintain, and have a high throughput capacity. Durability implies that the machine not be designed with a degree of sophistication so high as to require undue care in its operation and maintenance. The reason is that it is subjected to conditions far more rugged than those to which complex machines generally are exposed. Reliability implies both unremitting uniformity of performance and the unfailing fulfillment of the function for which the machine was designed. This latter is a difficult requirement since refuse is an extremely heterogeneous and variable material, and hence the selectivity range of the machine must be extremely broad and yet very precise in each selection. A high capacity would require that the response of the machine to code signals be swift (a matter of microseconds), otherwise hopelessly huge backlogs of unsorted refuse would accumulate at the sorting site.

Automated sorting systems fall into three broad classes, namely, dry processing, wet processing, and a combination of the two. In dry processing, as the name implies, the material being sorted is kept in a dry state throughout the process. On the other hand, in wet processing, the raw wastes are converted into a slurry, which is then subjected to further processing to remove the various components. In a combination system, the material is first subjected to dry classification and the residue, usually fibrous, is slurried and processed.

Each approach has its advantages and disadvantages. With the dry systems it is possible to take advantage of air classification and fluidized (air suspension) systems. The wet systems make use of centrifugation, settling, and filtration. On the other hand, wet systems carry with them the grave disadvantage of potential water pollution of the severest sort. As a consequence, wet separation systems must always include a wastewater treatment system. The combination systems, of course, share the advantages of the wet and dry approaches and are designed, it is hoped, to eliminate or minimize those steps which are accompanied by disadvantages of certain of the components in the systems. For example, only a small fraction of the wastes are slurried, and then only after most of the more objectionable contaminants have been removed. Consequently, the volume of the wastewater to be treated is small, and its quality is far less degraded than in wet processes.

A major disadvantage of all automated systems is the inability to make a sharp distinction between the various classes of components, e.g., paper and plastic film.

Mechanical sorting systems are based on physical, electrical, and electromagnetic properties of waste materials for detection and separation. Magnetic separation of ferrous metals probably is one of the oldest of the nonmanual salvage techniques. In operation, a conveyor belt moves the refuse to be processed past a magnetic belt which removes practically all of the ferrous metals. In refuse management, magnetic separation is used mostly to remove metal food containers ("tin cans") from the waste stream.

Another rather old system is based on inertial separation. Three versions of inertial separation are available: ballistics separation, the "Secator" system, and inclined conveyor separation. In ballistics separation, the material to be sorted is impacted by a heavily constructed rotating impeller and is flung horizontally at a slightly upward angle into a divided bin. The heavier and more resilient pieces leave the impeller at high speeds and are propelled to the far end of the bin. The less dense particles, having a greater degree of aerodynamic drag, are propelled a lesser distance and hence drop into that portion of the bin which is closer to the impeller.

In the "Secator" system, particles are projected by a high speed conveyor belt against a metal plate placed a short distance opposite the discharge end of the belt. The particles bounce off the plate and drop on the surface of a rotating drum from which they are propelled to varying distances. From the description, it becomes apparent that the separatory function of the machine depends upon variations in elasticity and densities of the material passed through it. The more resilient particles bounce further back from the metal plate. The action of the drum is based on density. Heavy and resilient materials fall off one side of the drum, while the light and inelastic material is discharged on the opposite side. The plate and drum may be adjusted to bring about variations in separation.

The third inertial system may be termed "inclined conveyor separation." In this system the material is dropped from a conveyor belt onto an upward sloping steel plated conveyor. Upon falling on the sloping conveyor, the heavier particles bounce downward and fall into a bin, whereas the lighter ones are carried over the top and drop into another bin. Different modes of separation may be obtained by varying the angle of incline of the steel-plate conveyor.

Other systems depend upon conductivity, photometry, radiometry, and X-ray detection. A difficulty in using such systems is that each individual particle in the waste stream must be analyzed, a process that could be quite slow. Such methods probably would be useful mainly in two-component systems (e.g., high and low conductivity, high and low X-ray attenuation) even though the detection may cover a wide, continuous band of readings. Conductivity detection involves the passage of particles through an electric field or in proximity to electric contacts. The pieces that are conductive (metals) are detected and diverted from the main stream. In photometric systems, the reflectance of a laser beam from the surface of the particles is measured, and separation is based on this measurement. Radioactive property of particles is the basis for separation in radiometric separation. The degree of attenuation of X rays passing through a particle is the basis for separation in the X-ray attenuation method.

A system has been developed that depends upon optical properties for separating colored from clear glass. In operation, screened pieces of mixed (colored and clear) glass are dropped from the end of a conveyor belt and fall past a light, a photocell, and an air jet arranged in the order named. The light source and photocell are positioned such that light is reflected from colored glass to the photocell, whereas it simply passes through the clear (transparent) glass. Transparent glass then falls directly from the belt into the bin directly below the end of the belt.

A method that has proved quite effective in separating certain components from the waste stream is dry fluidized bed separation. This technique was developed in Britain. A dry, uniformly granulated medium, usually of metal, is "fluidized" on a vibrating table by air blown upward through holes in the table. About one-third of the table is inclined upward at an angle of 5 to 10°. The material to be separated is granulated to a size larger or smaller than the fluidized medium and is dropped on the table. Lighter material "floats" on the medium and is moved along the horizontal portion of the table and dropped into a bin. Fluidized material deposited with the material is screened and recycled to the fluidized bed. Heavy material "sinks" and is carried up the inclined part of the table and falls into another bin. It, too, is screened and the medium is recirculated to the bed.

A system which has become a part of practically all dry separation schemes for municipal refuse is the air classification method.[35,36] In the system a zig-zag or other configuration tube of rectangular cross section is oriented vertically, and air is blown or drawn upward through the tube. Dry, granulated or shredded materials are introduced into the tube about midway between top to bottom. Particles of a density and drag not permitting support by the moving air fall down and out of the bottom of the tube. Lighter particles with more drag are caught up in the air flow and pass out of the top of the tube. The system is used mostly to accomplish a preliminary separation into "lights" and "heavies." The "lights" include most of the fibrous materials and film plastic. The "heavies" consist mostly of metals and heavy organic material (garbage, garden debris, etc.). Both outputs are subjected to further screening and other separation processes.

The presently existing wet systems, the "hydraposal" and "Fiberclaim" systems, were developed by the Black Clawson Co. to process municipal wastes.[37] The basic concept for both systems is that of introducing the whole waste stream into a "hydro-pulper" in which all reducible materials are broken, ground, and pulped into a slurry. Large metallic objects are ejected from the hydropulper and go through a "junk remover." The pulped slurry is then dewatered and reduced in a fluid-bed reactor to an inert residue for landfill; or it is further screened, cleaned, and dewatered to recover the paper fiber. Thus, it is theoretically a complete system. Among the more serious problems that have beset the system are very severe water pollution and a low level of effectiveness in separating refuse components. At the time of this writing, paper fiber recovery was on the order of 35 to 40 percent, and the recovered fiber was of very poor quality. Another problem needing solution was the separation of aluminum bottle caps, etc., from glass.

Aluminum and plastic film pose special problems in automatic sorting — the former because of its nonmagnetic properties and the latter because it is light in weight and is comparable to paper in resistance to air. As a result of these two characteristics, it is practically impossible to separate the two (paper and plastic) by inertial separation, air classification, or screening. The nature of the problem with the removal of aluminum is largely one of frustration due to the difficulty of recovering a monetarily valuable resource. The nature of the problem resulting from the difficulty in separating plastic film from paper is the interference from the presence of plastic with the recovery of paper fiber and the consequent lowering of the quality of the latter. Plastics also detract from the quality of the compost product in those resource recovery operations that involve composting. Some progress apparently has been made in aluminum separation in that an aluminum sorting device has been designed which depends upon electromagnetic phenomena.[36] A charge is imparted to the aluminum particles as they are passed through an electric field. Thereafter, upon passage through an electromagnetic field, they are repulsed into a collecting chamber. Other systems involve flotation. Shredded wastes are placed in a succession of media, each of which has a different specific gravity. The components are separated by reason of differences in density, in that density matches specific gravity.[38]

A process for separating plastic film from paper (the "Cal

system") has been developed at the University of California Richmond Field Station as a part of a total system by Trezek and his associates.[39] A description of the total system is given herein, albeit briefly, because it is a good model of a process that has been proved to be technically and economically feasible on a meaningful scale, i.e., capacity of 4 tons municipal refuse per hour. A flow diagram of the system is presented in Figure 3.

The Cal system begins with a front-end dry-processing step. In this step milled refuse is passed through a vertical air-classifier, thereby separating the incoming waste stream into a light and a heavy fraction ("lights" and "heavies"). The heavy fraction, now practically devoid of fiber, is further separated into its components (ferrous and nonferrous metals, glass, and organic and miscellaneous compounds) by passage through specially designed trommel screens and fluidized bed screens. The organic rejects produced in this step are suitable for composting or for anaerobic digestion. The "lights" consist predominantly of fibers contaminated with plastics and with fines composed principally of ground glass, dirt, some fine organic materials, etc. The contaminants are removed by screening. (Discarded fibers from the "lights" fraction can be composted or digested along with the rejects from the "heavies" fraction.)

The "accepts" from the screened light fraction can be processed in three ways: (1) The material can be used as a fuel supplement or as raw material in a paper mill equipped for processing secondary fibers contaminated with plastic, hot metal, etc. This material deteriorates extremely slowly when stored because practically all of the putrescible material has been removed. (2) The screened material can be milled again and used as a fuel for certain types of combustion applications or for pyrolysis. (3) The material can be further processed to upgrade the fibers, which is accomplished by wet processing. By postponing wet processing to this step only a small fraction — and that the "cleanest" fraction — is converted into a slurry. Hence, degree of water pollution and amount of water to be cleaned are far less than with a completely wet processing system.

The wet phase of the Cal system is initiated by pulping the screened "lights" to a consistency of approximately 3 percent. The design of the pulping equipment is such that the paper

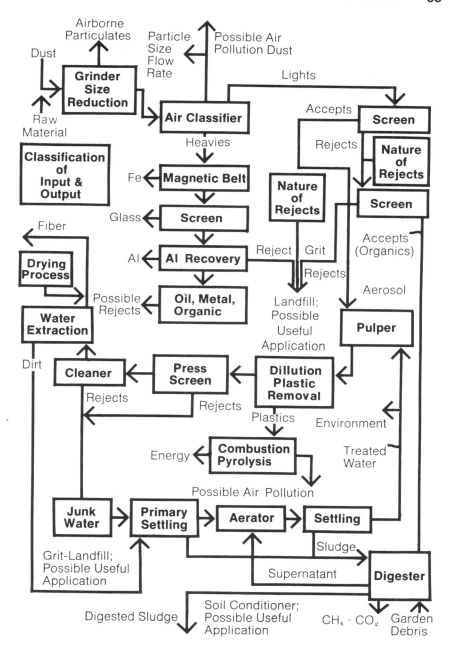

Figure 3: Flow diagram of the Cal resource recovery system

component of the "lights" is macerated into fibers, while the dimensions of the plastic particles remain unchanged. The slurry is then diluted to a consistency of about 0.5 percent is an operation in which the plastics are removed. The fiber slurry is passed into a cyclone and cleaned by a two-step centrifugation process. Fine glass, grit, and dirt contaminants that escaped the preceding removal steps, as well as fiber bundles and some useable fibers, are removed in the first step. These rejects exit through the bottom of the cyclone. Hot metal, wax, and some of the remaining fine plastics are discharged from the center probe of the cyclone. The reject stream from the bottom of the cyclone is further treated in a reject cleaner to reclaim some of the useable fibers contained in it. The rejects from the reject cleaner are discharged from the system. Despite the second pass, about 5 percent of the total fibers that entered the wet-processing phase are lost. The "accepts" from the reject cleaner are combined with the first cleaner "accepts" and passed through a second cyclone for a centrifugation step. Because the reject stream from the center probe and bottom in the second steps are much less contaminated than those from the first step, they are returned to the beginning of the first step and are reprocessed. The interposition of a holding period in a storage tank between the first and second steps leads to a cleaner fiber. The fiber produced in the two steps is suitable for media and liner board manufacture. A third cleaning step would produce a very high-grade fiber. The processed slurry is dewatered, and the fibers are dried and baled for shipment. The water is recycled through the system.

Technology for the Composting Step

■ **Windrow — Shape and Dimensions:** As stated earlier, in windrow (open) composting, the wastes are stacked in piles immediately after the completion of any required preprocessing steps. The piles may be conical in shape or elongated to form windrows. In cross section, the piles may range from trapezoidal to triangular. Generally to be preferred is a roughly dome-shaped cross section, with the sides tending toward the vertical, i.e., slanted at about a 70 to 80° angle. The reason for rounding the top of the pile is to promote the shedding of rain during wet periods or to prevent the accumulation of snow in the winter. The melting snow could result in the pile being flooded. Where rainfall is not sufficient to saturate a pile, the conical shape would not be imperative.

The key factor to keep in mind when determining the dimensions of a pile is the need to maintain aerobic conditions in it. Since, as pointed out earlier, the major source of oxygen is the air trapped in the voids between the particles of wastes, the dimensions of the pile should be such that the volume of the individual interstices is not overly reduced by the crushing of the material. Obviously, the higher the pile, the more intense the crushing force, and hence the greater the diminution of interstitial volume. Since the potential reduction of interstitial volume is partly a function of the "structural" strength of the material being composted, permissible height of a pile is determined by the nature (*sic* "structural strength") of the waste. A pile of wastes having a large content of straw, wood chips, sawdust, rice hulls, etc., could be much higher than one having a large concentration of paper.

Another factor that could place an upper limit on the permissible height of a pile could be one which may stem from the turning step — i.e., from limitations of the worker or of the turning machine. If the turning is to be done manually, the height of the pile should be approximate to that of the average worker. The type of machine determines the height of the pile it can handle. Experience gained by this author indicates a height of 5 to 6 ft. as far as maintenance of the integrity of the interstices is concerned. The width may be one determined by convenience or by the one naturally assumed by material stacked to form a 5- to 6-ft.-high pile. If the material is turned mechanically, the width suited to the turning machine is the one to have. Generally, a width of 7 to 8 ft. seems to be satisfactory.

• *Aeration* — The most effective way to aerate a pile of composting refuse is to turn it. Turning results in a reconstitution of interstices with the entrapment of fresh air to serve as a renewed supply of oxygen. Turning is accomplished by tearing down a pile and then reconstructing it. The reconstructed pile may be on the original site or immediately adjacent to the original. Regardless of the method of turning, in reconstructing a pile, the material should receive a tumbling or "trickling" action rather than a compressive one. A tumbling action results in a "fluffed" up pile that has a high degree of porosity. Moreover, if it is at all feasible, the material forming the outside layers of the reconstructed pile should come from the interior of the original pile, as indicated by the sketch in Figure 4. The reasons for this reversal of layers are obvious: (1) to ensure uniformity of decomposition, since composting proceeds more rapidly in the interior of the pile; and (2) to expose at some time all of the material to the high temperatures prevailing inside the pile. The need for this is discussed in the section on public health aspects. (Note that neither of the two reasons has anything directly to do with aeration, although aeration does affect the phenomena involved.)

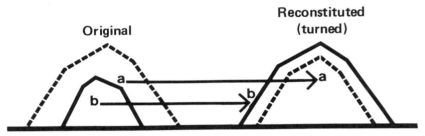

Figure 4. Rearrangement of layers in turning a compost pile

When the reversal of layers cannot be accomplished, its lack can be compensated for by stepping up the frequency of turning. This latter increases the chances of a given aliquot of material being exposed to the interior of the pile. Moreover, while required frequency of turning is chiefly a function of oxygen needs, certain other factors because of their bearing upon oxygen availability also exert a determining influence. The precise bearing between these factors and aeration was described earlier. For example, if the moisture content of the pile is excessively high, more frequent turning is indicated. On the other hand, if the moisture content is satisfactory and the pile has the desired degree of porosity, frequency of turning can be less. Another factor influencing frequency is the collective needs of the operation. When space requirements are not critical, and hence a more leisurely rate of composting is permissible, then frequency can be reduced drastically, provided the public health aspects are satisfied.

For a typical, normal refuse having a maximum moisture of 55 to 60 percent, and an operation in which the time allotted to composting must be kept at a minimum, the refuse need not be turned until the third day after grinding. Thereafter it should be turned every other day for a total of four to five turns. The program is illustrated in Figure 5. In the normal course of events, by the fourth or fifth turning the material will have been so advanced toward stabilization as to require no further turning. More frequent turning would do little to hasten the process unless the material were very wet or compacted. When time is not of the essence, once-a-week turning is sufficient.

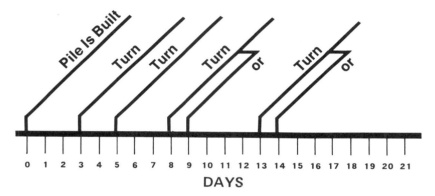

Figure 5: Turning program for rapid composting

Equipment for turning can range from the humble pitchfork used in manual turning to a complex "automatic" turning machine. Obviously, manual turning is economically feasible only where labor is inexpensive and plentiful. For estimating manpower needs, one man can be assumed to be able to turn from 4 to 6 tons of material per 8-hour day.

For years, versions of automatic turning machines have been utilized by commercial growers to compost the manure used in mushroom production. Typically, a machine consists of a tractor equipped with a rotary scoop in the front. The scoop collects the material and drops it on an endless belt that passes over the cab of the tractor. The belt drops the material into a mobile form attached to or towed by the tractor. In action, the machine works into a pile head-on, reforming the pile as it moves through the length of the pile.

A machine designed specifically for large-scale turning in municipal compost operations is the Terex 74-51 Composter manufactured by the General Motors Corp. According to the manufacturer, in a test run the machine turned about 800 tons (726 metric tons) of municipal refuse per hour.[40] In the run, the refuse had been stacked in rows of 100 tons (91 metric) per row. Each row averaged about 400 ft. (122 meters) in length and about 14 ft. (4.25 meters) in width. At such a turning rate, one machine would suffice for a 320-ton/8-hour day (290 metric tons/8 hours) operation in which the detention period is 30 days. A photograph of the machine is shown in Figure 6. As shown in the picture, the machine is equipped with an endless belt of paddles or slats, which lifts and aerates the material, and a blade, which diverts and forms

Figure 6: The Terex 74-51 composter

the new pile. The new pile is formed alongside the old one. For design purposes, a space the width of the machine should be allowed between the pile to be torn down and the new one to be built, as illustrated in Figures 7 and 8.

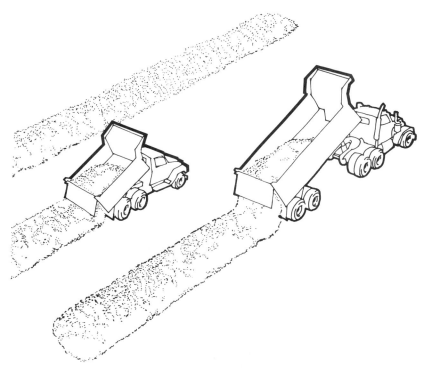

Figure 7: Setting up the windrows for the Terex 74-51

A front-end loader can be used to turn the windrows, but it is a poor substitute for a specially designed turning machine. It has two disadvantages: (1) a tendency to compact the composting material and (2) inefficiency. For example, in the compost operation at Mexico City, a front-end loader with a 3.5-cu.-ft. bucket can turn only 200 to 300 tons (180 to 270 metric tons) in an 8-hour day. Moreover, the cost of such a loader is about twice that of the Terex 74-51.

Methods of aeration other than turning involve designing processes that fit neither the classification "windrow" nor

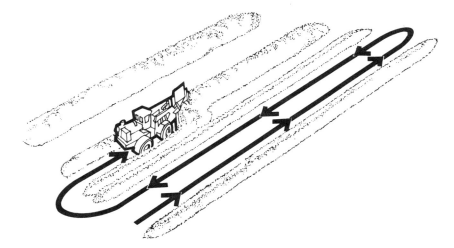

Figure 8: Arrangement for turning windrows with the Terex 74-51

"mechanical." They may involve some structural containment, which removes them from the windrow category; and yet they are not sufficiently enclosed or mechanized to be termed "mechanical." While this may be true, these methods nevertheless are discussed in the section on windrow composting because of clarity and ease of presentation and because they represent only a moderate departure from the true windrow.

An example is the method of forced aeration. In one such approach, the material is placed over a perforated floor through which air is forced. To promote a uniform distribution and maximum effect from the entire input of air, the sides of the composting mass may be enclosed. In other words, the material is placed in a long bin. At first sight, forced aeration seems to be an attractively simple and inexpensive, yet effective, method of aeration; and it may be so for certain types of wastes. However, it has not proved to be particularly successful with municipal refuse. The problem is that the streams of forced air coalesce to form channels through the mass of refuse, thereby short-circuiting the air out of the pile. Material a few centimeters from the channels do not receive enough oxygen, and the pile becomes largely anaerobic. Moreover, the movement of air through the channels dries the area immediately adjacent to them. Forced aeration has its best potential with highly porous masses of material — e.g., material in which straw, sawdust, or rice hulls constitute the predominant carbonaceous material.

Apparently forced aeration can also be used with partially dried cattle manure. For example, Senn[41] found that forced aeration worked well with composting dairy manure, provided the moisture content of the mass was at a maximum of 50 to 60 percent, i.e., when the manure was not wet enough to compact. Even then he found some turning to be mandatory. The installation of chimneys in the bin or windrows probably is more ineffectual than forced aeration. If they function at all, which is dubious, they would act to channel air out of the pile rather than into it, because an outward draft would be created because of the difference between the temperature in the pile from that of the ambient air.

Another approach calls for placing the wastes in bins fashioned or constructed of hardware cloth.[42] Theoretically, oxygen comes through the diffusion of surrounding air into the mass contained in the bins. However, the amount of the diffusion into composting material is minute if any. The insulating (heat retention) properties of a pile of milled wastes is evidence of the lack of air movement into or out of a pile. A second major problem arises in a large-scale operation, namely, impracticality. The impracticality is due (1) to the inability of hardware cloth to withstand the rough handling to be expected in a large-scale operation and (2) to the excessive amount of handling because of the necessarily small bins; hence a large number of bins is required in any municipal-size operation.

A final method, developed for composting cannery wastes, is even more difficult to classify either as open or as mechanical because it has in equal number characteristics that are found in open and in mechanical systems.[17] With this method, the material to be composted is placed in an elongated outdoor bin. Mixing is accomplished by passing a mobile endless belt through the composting material. The belt is equipped with slats placed transversely along its length. As the belt moves through the mass, it picks up material in front of it and deposits it in back, thus thoroughly stirring and agitating the material. Strictly speaking, the only characteristic that justifies to some extent the placing of this system in the windrow or open category is that the bins are not enclosed in a shelter. On the other hand, as shall be seen later, there is a mechanized method in which a traveling endless belt is used.

Another process that should be mentioned at least briefly is the Bühler process. Specific mention of it is made because it exemplifies an occasional tendency of a manufacturer of a grinder to designate a conventional compost process by a label based upon that of his hammermill. Thus, the Bühler system is essentially a windrow compost system in which a Bühler grinder is used. The compost plants in Mexico City, Guadalajara, and Monterrey (Mexico) utilize the so-called Bühler system.

Aside from the volume of the wastes to be treated, the area requirement for a *bona fide* windrow process is a function ultimately of the speed at which the material is composted. With a combination of optimum environmental and operating conditions (especially frequent turning) and a readily composted waste, such as manure, garden debris, vegetable trimmings, and wastes having a highly putrescible content, the material should be sufficiently composted, i.e., stabilized, within a 14- to 16-day period. In this author's studies an unmilled mixture of leaves, garden debris, and vegetable trimmings was composted to stability within 12 days. In general, the higher the C/N ratio, the longer will be the period required. A safe allowance for high-rate composting of municipal refuse having a C/N ratio lower than 30:1 would be about 30 days. Under these conditions (3 lbs./person/day, density in windrows at 400 lbs./cu. yd. and piles 6 ft. high [1.4 kg, 251 kg/m^3, 1.8 m high]) the estimated area requirement for a windrow system would be from eight to 10 acres (3.2 to 4 hectares)/100,000 people. This allows for a 100 percent safety factor. Of the eight to 10 acres, between six and seven acres would be for composting, and two to three acres for building, storage, and general yard area.

■ **Site Preparation and Shelter:** During the active period, i.e., when the material is undergoing a regular program of turning, the operation should be conducted on a paved surface (e.g., asphalt of parking lot specifications). The reasons are (1) to ensure maneuverability of equipment during rainy periods, (2) to facilitate collection of leachate if any should be formed, (3) to prevent fly larvae from escaping the pile by burrowing into the ground, and (4) to promote general "good housekeeping." For a high-rate process, this would mean a hard surface for about the first two weeks of composting. While not essential, a roofing over the windrows during the first couple of weeks would be advisable, especially if the region is subject to snowfall.

■ **Applicability of Windrow Systems:** At one time it was thought that windrow composting would be suitable only for small operations or where extensive land areas are available. The reasoning rested upon the fact that with windrow systems, raw refuse is exposed to the elements until composting has advanced far enough to destroy the characteristics of raw refuse. Hence, the site of operation might be a focal point for vector and rodent generation and sustenance, as well as a source of objectionable odors. The larger land area requirement was thought to be a logical corollary of a process expected to be slow. However, a combination of the development of an excellent turning technology and a better understanding and hence application of operating conditions has shortened the compost time to at least the equivalent of any existing mechanical system. Moreover, these improved operational procedures have resulted in a minimization of vector, rodent, and nuisance problems. On the other hand, it can be pointed out that every municipal-scale mechanized ("enclosed") compost operation in the United States was at one time or another beset with odor problems — so much so that some were threatened with legal action by regulatory agencies.

Mechanical

As stated earlier, mechanical systems are designed to provide optimum conditions, and hence accelerate the process. An additional intention is to isolate the wastes from the environment and thereby eliminate an unfavorable environmental impact. (The unit that constitutes the realization of the design usually is termed an "aerobic digester" or simply a "digester.") An example of control of environmental conditions in a digester is the provision for adding water to the digester contents. This usually is done by installing a row of nozzles inside the unit. Other additions (e.g., lime, inorganic nitrogen, etc.) usually are made at the time the milled material is introduced into the unit. The major differences in design among mechanical digesters are in the method of aeration. Some provide aeration by tumbling or dropping the material from one floor to the next. Others call for stirring devices which rotate through the composting mass. Tumbling in a rotating cylinder is yet another approach. Finally, there is the combination of forced bottom aeration and stirring by means of a traveling endless belt. Of course, all of the systems have one feature in common: They are enclosed.

To better explain the various design approaches, examples of systems on the market shortly prior to or at the time of this writing are described herein. Needless to state, all rely upon the use of milled and sorted wastes.

Probably one of the more successful processes, although an expensive one, is the Dano process. In Figure 9 is diagrammed a typical layout of a Dano Bio-stabilizer plant. The process involves the use of a large, slowly rotating drum, the interior of which is equipped with vanes or baffles. Material is injected in one end and after one to three days of slow rotation is ejected from the opposite end of the machine. (The manufacturer recommends three days.) Aeration is accomplished by tumbling action. Air is injected into the interior of the drum to ensure a constant supply of oxygen. Following the sojourn in the drum, the contents are

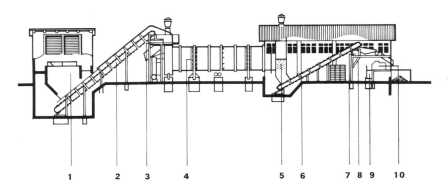

Figure 9: Typical layout of a Dano Bio-stabilizer plant. (1) receiving hopper, (2) conveyor, (3) magnet, (4) Dano-Bio-Stabilizer, (5) primary screen, (6) conveyor, (7) magnet, (8) vibrator screen, (9) conveyor for compost, (10) grinder

windrowed to permit "maturation" or "curing" over a period of one to two months — one month if the material is turned occasionally.

The Naturizer system is one in which the composting wastes are tumbled from one floor to the next. A Naturizer plant consists of two 3-floor structures. The ground material is transported to the top floor of the first structure by conveyor belt. It rests on the top floor for 24 hours, at the end of which it is dropped to the middle floor. This mode of action is made possible by constructing the floors as V-shaped troughs placed side-by-side and equipped such that they can be pivoted upside-down, and thus dump the material to the lower floor. The action is much like that with the shaker grates in an old-fashioned stove. The composting material remains over a 24-hour period on each of the two lower floors. After 24 hours residence time on the bottom floor, it is ground a second time and then transported to the top floor of the second structure in which the floor-to-floor dumping procedures are repeated. Thus, the retention time in the plant is six days. The processed material is then windrowed, and time is allowed for one to two months of maturation to take place. To ensure adequate "maturation" in that period, the windrows are turned occasionally.

A third system, the Metro system, makes use of forced bottom aeration coupled with stirring by means of a traveling endless belt, such as is diagrammed in Figure 10. In operation, ground refuse is dumped into elongated bins. Each bin has a perforated bottom, over the length of which is placed a set of rails on which the mobile belt moves. As soon as the bin is filled, the air stream is turned on. The belt is passed through the bin contents usually once each day. The recommended detention time seems to vary with the individual promoting the sale of the machine, and thus it may be from one to six days. Thereafter, it is placed in windrows to cure.

A system more complex in some respects is the Fairfield-Hardy unit. It involves the placing of ground refuse in an open cylindrical tank equipped with a set of screws ("augurs" or "drills") supported by a bridge attached to a central pivoting structure. The bridge rotates much like the arms of a lawn sprinkler. The arrangement is indicated by the diagram in Figure

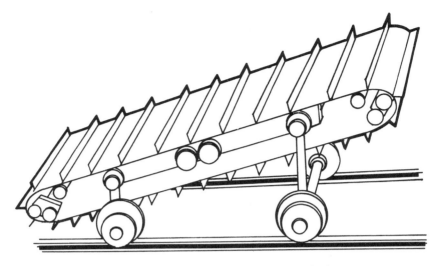

Figure 10: Traveling endless belt stirring device

11. The screws are turned as the arms rotate. The screws are hollow and perforated at their edges. Air is forced through the perforations and into the composting material as the screws are forced through the wastes. The detention period in the unit varies

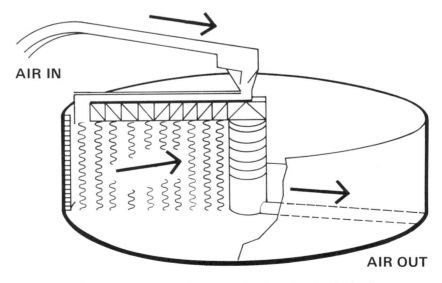

AIR IN

AIR OUT

Figure 11: Diagram of aeration device in the Fairfield-Hardy digester

and usually depends upon the inclination of the operator. If the detention period is short, the material is windrowed to permit maturation.

The last type of digester to be described is one that consists of a vertical silo divided into six (more or less) tiers. Each floor has an opening at its center to allow for a shaft that extends the height of the silo. The shaft is equipped with arms to which are attached plows. The axle is rotated and thereby moves the plows through the refuse on each deck. The material is dropped from deck to deck at a rate to provide a retention period of about one day per deck. A diagrammatic sketch of such a unit is shown in Figure 12. Although the entrepreneur claims that the material is composted by the time it is discharged from the bottom floor, it does require a considerable amount of further maturation. Two of the major problems inherent in the unit are a high power requirement and a tendency of the shaft to shear because of the high degree of torque.

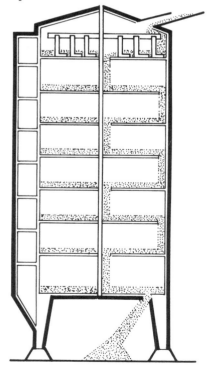

Figure 12: Diagrammatic sketch of the Earp-Thomas digester

■ **Relative Merits of Mechanical vs Windrow Systems:** A perusal of claims of the various vendors of mechanized systems will show that they (the claims) have one thing in common, namely, a very brief compost time requirement, mainly on the order of two to about six days. It also will show that they all specify a one- to three-month maturation period.

The reality is that the respective units do not compost the material within the time specified by the promoters. The brief retention time within the machine corresponds to an intensive initial aeration and perhaps the decomposition of the more putrescible components. If this so-called composted material were to be stored or sacked (unless thoroughly dried to less than 12 percent moisture), it would give rise to very severe nuisances — hence, the prescription of an extended maturation period, because it is in this period that composting is completed. The upshot, therefore, is that total composting is about the same with both the windrow and the mechanical approaches. It is for this reason that the sobriquet "high-rate composting" is not given to mechanical composting in this text. The question is whether or not the extent of the composting accomplished in the machines warrants the huge capital and operating costs expended to attain it. The doubt becomes especially pervasive when one considers that the *total* compost time with the far less expensive windrow system is the same as with a costly mechanical (enclosed) system.

A system that can be operated on a continuous basis is a potential advantage to a large-scale application. A continuous system is one in which input continues uninterruptedly throughout the 24-hr. day and from day-to-day; and of course, the output correspondingly matches, unless a part of the material is destroyed or lost during the operation. With a batch system, an amount of material is introduced into the system at one time and is held and treated and then discharged only after the entire batch has been processed. A continuous process generally is the desideratum for large-scale operations because of lower capacity requirements in terms of unit size and of operation. Both windrow and mechanical processes can be operated as continuous systems, at least in the broad sense of the term. A windrow can be operated such that new material is continuously added at one end of the row and completed material taken from the other end. Of course this example need not be taken literally, since the windrows can be arranged serially according to the age of the pile. The important point is that the new material can be introduced throughout the day into the operation as a whole and also withdrawn. Strictly speaking, the designation is less readily applicable to mechanical systems that have compartmentalized units. In such units, the raw material is placed *en masse* in a compartment, held there for a specific period of time, and then discharged into a second compartment, and so on. However, the deviation is such that the benefits of a truly continuous system are attained.

It is in the matter of capital and operational requirements that the mechanical systems suffer most by comparison with windrow systems. Inasmuch as actual monetary costs are treated in detail in a later section, at this time only generalities are stated, and that mostly for the sake of comparison. In making the comparison no mention is made of preparation (sorting, grinding, etc.) because the requirements for both approaches are precisely the same. It is stressing the obvious to point out that the capital outlay for a mechanized system will far surpass that for a windrow system. At the most, the capital outlay for a windrow system would be for hard-surfacing a small part of the land area, and perhaps roofing an even smaller part, and for a mechanical turning machine. (Both types of systems call for screening or grinding to prepare the composted material for storage or utilization.) On the other hand, even the simplest of mechanical systems involves the use of a carefully designed structure or piece of equipment on a scale large enough to contain and process at least one-day's and more usually, three- to six-days' output of wastes.

The result is that capital costs are in terms of four-digit numbers per ton with windrow composting as against five-digit numbers with mechanical systems. The disparity becomes even greater with respect to operating costs. Because of the extent and sophistication of the equipment employed in a mechanical system, its power and manpower requirements exceed those for windrow systems. The power needed for a windrow operation is at most the modest amount used in operating one or two mechanical turners. On the other hand, a mechanical system must utilize energy to light and heat the control room, to monitor the digester, to heat it when needed, and to power the moving parts of the digester. For example, the Dano system utilizes from 12 to 15 kwh. to process a ton of refuse.[34] On the other hand, the reported power expenditure for a plant processing a ton of sludge (wet weight) in a 600-ton (540 metric ton)/day windrow operation was less than 4 kwh./ton plus about $0.18/ton for fuel and oil to operate a turning machine.[43]

Because of the absence of complex machinery and the consequent need to maintain and supervise such machinery, the skilled manpower required to run a windrow operation is much less than that with a mechanical system. The reasons are not hard to find. The promoter of a particular system naturally wishes to quote a low cost per ton to the potential buyer, especially in comparison with that of competing systems. To do this, he all too often makes unrealistically low estimates of any requirement that adds to the cost per ton. The penalty for the underestimation often is first manifested by the emanation of the foul odors that betoken improperly composting material.

Although the promoters of mechanized systems point to low land requirements, e.g., 2 acres (0.8 hectares)/100,000 people, the fact of the matter is that land requirements for both systems are identical because both involve windrowing for comparable lengths of time.

Final Processing

Before the compost product can be utilized or placed on the market, it generally must be subjected to a final processing. The major purpose of this step is to eliminate materials and oversize particles that would detract from the quality and, hence, the utility of the product. The final processing may come either before or after storage. Carrying on the step after storage has the advantage of allowing further decomposition to take place. This is especially true when municipal refuse is composted. The further along the decomposition has progressed, the easier it is to accomplish this step.

The processing may consist of one or a combination of two operations, namely, screening and/or a second grinding. The energy requirement for the final grinding is perhaps 30 to 50 percent lower than that for size-reducing the raw refuse because the material has become more fragile and consequently easier to mill. Moreover, the quantity to be milled is much less because noncompostable material has been removed and some of the volatile matter has been destroyed. However, in most cases a second grinding is not needed, inasmuch as screening alone suffices. During the compost process step and storage, an appreciable amount of particle size reduction is brought about because the handling results in particles being rubbed against each other, and thus gradually being abraded. This fact together with the increase in fragility mentioned in the previous paragraph leads to the breaking up of the particles. Oversize particles retained on the screen are either recycled within the operation or are landfilled.

Difficulties are likely to be encountered if a horizontal shaker screen is used. A major one usually is in the form of "blinding" of the screen perforations by oversize particles or by plastic film. The problem is aggravated if the material has a high moisture content. Trezek and his fellow researchers have found that the "blinding" can be avoided by the use of a trommel screen.[39] Blinding is hindered by the tumbling action imparted by the screen. In the Mexico City operation, the material is partly fluidized on a horizontal shaker screen by forcing air through the screen.

USE OF THE PRODUCT

Preliminary Remarks

A discussion on the use of compost usually falls into three broad aspects, i.e. along three lines: (1) the fertilizer value of the product, (2) its utility as a soil conditioner, and (3) the use of the product to reclaim land. Of the three potential uses, the second is the most important and has received the most attention. For that reason it is emphasized in the discussion that follows. As is to be expected, the relative values for each of the characteristics depend upon the nature of the material and the manner in which it was composted.

Fertilizer Value

The nitrogen, phosphorus, and potassium (NPK) content of compost from refuse in the United States generally has been quite low. The reason becomes clear when one considers the composition of average U. S. refuse. Although about 40 percent (±10 percent) theoretically is compostable, the C/N ratio of this fraction is too high to permit the production of a compost with a satisfactorily low C/N ratio, unless nitrogen is added from another source or a portion of the degradable carbon is removed, as for example in a resource recovery operation. Of course, this would not be the case if sewage sludge were added to the refuse, nor with wastes in less industrialized countries. Refuse produced in the less industrialized countries is rich in organic matter and hence has a typically low C/N ratio. Composted U. S. refuse generally had a nitrogen content less than 1 percent and a phosphorus and potassium concentration of less than 1 percent each. (The historical tense "had" is used because at the time of this writing no municipal-size plant was in operation in the United States.)

The product from the three Mexican plants in operation in 1975 ranged from 1.2 percent to 2 percent in concentration each of nitrogen and phosphorus and less for potassium. The NPK for composted manures depends upon the animal source of the waste. Composted cow and horse manures usually have a nitrogen content of about 2 percent or less, and a comparable phosphorus content. Composted poultry manure may have a nitrogen content as high as 4 to 5 percent; and composted sheep and pig manure, from 2 to 3 percent.

From these numbers it can be seen that regardless of type of waste used as raw material, compost by itself generally will not have an NPK value sufficient to meet the legal specifications for fertilizer, namely, 6 percent. Therefore, if it were planned to advertise a compost product as a fertilizer, it would be necessary to enrich it with inorganic nitrogen, phosphorus, or potassium.

A point to remember in comparing required loadings to satisfy the fertilizer needs of a crop is that the fertilizing elements of compost are more efficiently used than those of inorganic fertilizers. A prime reason is that unless a field is overloaded with compost, none of the fertilizer elements of the compost are leached away; hence, all eventually are available to the crop. On the other hand, as much as 30 to 35 percent of inorganic nitrogen and 10 to 20 percent of phosphorus is lost as leachate;[44] thus, this factor must be taken into consideration when estimating inorganic fertilizer loadings. The compost supplies the carbon used by the microbes to immobilize the nitrogen into the form of cellular mass. The critical shortage of energy sources, especially of natural gas, has added a new factor, energy, to consider with respect to the nutrient aspects of compost. The use of organic fertilizer (including compost) instead of chemical fertilizer can result in a two-thirds energy saving, i.e., 6,000 Btu's/dollar value of crop with organics, as against 18,400 Btu's/dollar value with chemical fertilizer.[44] (N.B.: These numbers are for manure. The savings may be less for compost, since it is necessary to consider the energy invested in making the compost.)

A very useful characteristic of compost is its possession of a full complement of trace elements. Inorganic fertilizers on the market generally are lacking in these elements, to the detriment of the crop and of the consumers of the crop, since the major source of the elements for animals is through their food intake. A note of caution should be borne in mind to the effect that overloading a field may result in an overdosage of one or more trace elements and consequent damage to the crop.[45]

Soil Conditioner Aspects

Compost finds its greatest and most valuable use as a soil conditioner. Thus, while compost may come out unfavorably in a comparison with inorganic fertilizer in terms of respective monetary costs for equal amounts of fertilizer elements, the balance is tipped decidedly in favor of compost when overall utility in crop production is concerned. The superiority of compost in terms of crop production is due to its soil-conditioning capacity. It is this characteristic that leads to a more efficient use of fertilizer elements. Furthermore, this characteristic makes it appropriate to designate finished compost by the general term *humus.*

Although the term *humus* is not subject to rigid definition, it may be described as a complex aggregate of amorphous substances resulting from the microbiological activity in the breakdown of plant and animal residues.[46] Chemically, humus is a heterogeneous material that includes various compounds synthesized by microorganisms, complexes resulting from decomposition, and plant material resistant to further breakdown. The principal constituents may be classified as derivatives of lignins, proteins, certain hemicelluloses, and cellulose. Humus is not in a biochemically static condition. In the presence of suitable environmental conditions, activities of microorganisms continue to bring about changes in its composition until it is eventually oxidized to inorganic salts, CO_2, and water. It has a high capacity for base exchange, for combining with inorganic soil constituents, and for water absorption with consequent swelling.

When used in the soil, compost in its role as humus has many characteristics which exert a beneficial influence on the soil itself and on growing vegetation. Organic acids resulting from the metabolic breakdown of organic matter form a complex with inorganic phosphates in the soil. In this form phosphorus is more readily available to higher plants. Both phosphorus and nitrogen are involved in a storing effect peculiar to humus. The precipitation of phosphorus by calcium is inhibited; and nitrogen, being converted to bacterial protoplasm, is rendered insoluble. In effect, it is stored. This nitrogen again becomes available when the bacteria die and decompose. The effect is to prevent leaching of soluble nitrogen and to make it available at a rate at which it can be utilized by plants. The gradual decomposition of insoluble organic matter results in the continuous liberation of ammonia, which is then oxidized to nitrites and nitrates.[47] The "binding" of nitrogen in the form of microbial protoplasm has the effect of preventing nitrification (mineralization) of nitrogen introduced into the soil.[45,48,49] This is an advantage is terms of preservation of ground water quality because it is in the nitrate form that nitrogen is most soluble.

Often more important than its nutrient effects are the physical effects of humus on soil, because soil structure may be as important to fertility as nutrient content. Compost promotes a soil aggregation or a tendency to "crumb," which in turn enhances the air-water relationship of the soil. Thus the water retention capacity of a soil is increased and a more extensive development of the root systems of higher plants is encouraged. The aggregation is brought about by cellulose esters (cellulose acetate, methyl cellulose, and carbomethyl cellulose) formed in the course of bacterial metabolism.[50] Other benefits accruing from bacterial metabolism of the compost are an increased ability of the soil to absorb rapid changes in acidity and alkalinity (increased buffering capacity) and neutralization of certain toxic substances.

The manifestation of these and other beneficial effects in the form of enhancement of crop yields is amply documented in the literature on composting.[45,51-53] A multitude of other references that could be cited are reviewed in the series of reviews published in past years by this author.[54-58]

■ **Constraints on the Use of the Product:** The urge to attribute unalloyed benefits to the use of compost that might arise from reading the preceding favorable account should be tempered by the realization that its use must be accompanied by certain constraints. These constraints pertain principally to the maturity of the product, to the amount of loading per acre of soil, and to public health precautions.

The effect of maturity comes primarily in the form of nitrogen "robbing" and other competitive manifestations. The competition is for nutrients and is between the soil microflora and the higher (crop) plants. Inasmuch as the microflora have a far larger surface to volume ratio than do the root hairs of the higher plants, they are in a better position to take in larger amounts of nutrients, and at a higher rate than is possible for the root hairs to do so. Moreover, because of a greater simplicity of morphology and cellular organization, the microflora can utilize the nutrients more efficiently and rapidly.

The upshot is that unless the available reservoir of nutrients is large enough to satisfy the needs of both groups, the microflora will flourish at the expense of the higher plants. This unfortunate chain of events occurs when "immature" (fresh) compost[59] or compost with an excessively high C/N ratio is applied to the soil. Fortunately the inhibitory effect can be alleviated by adding more nitrogen (or other needed nutrients) to the soil. Moreover, the effect will have disappeared by the second year, and in fact crop yield in that year will be even larger than would be expected on the basis of amount of fertilizer applied. The reason for the enhanced yield in the second year is that the nutrient added in the first year was not lost to the soil. On the contrary, it had been stored in the form of bacterial cells and was released when the organisms died and were decomposed before the advent of the second year.

The second constraint arises from the presence of trace elements and certain metals in compost, especially composted refuse and sewage sludge mixtures. Since the trace and metal elements are present in minute concentrations, large amounts of compost may be added before effects harmful to the crop result. (Their relation to public health is discussed in the section on that subject.) Probably the main offender as far as composted refuse is concerned is boron.[5,45,60,61] Generally, loadings up to 40 to 50 tons per acre have been found not to be excessive, especially when added in the spring. As with nutrient deficiencies, but with the exception of heavy metals, harmful effects from an overload disappear with the passing of time.

■ **Public Health Aspects:** The public health aspects of compost pertain to: (1) the nature of the wastes (especially the source) to be composted, (2) the manner in which they were composted, and (3) the use of the product. An important aspect of the latter is the type of crop grown in the presence of the compost. Generally, the public health aspects of composting are positive in that they are conducive to the preservation of public health, and hence are usually treated as a separate subject under the general heading of "benefits." However, the decision to classify public health aspects as a constraint or as a benefit depends upon one's viewpoint — somewhat analogous to regarding a tumbler of water as being half full or half empty.

Public health aspects must be considered when excretory wastes of animals and man are composted. The aspects pertain to the possible presence of viable pathogenic organisms in the compost product. Heavy metals and toxic materials acquire a potential public health significance when industrial wastes are composted and, to a much less extent, when municipal refuse is the source. The literature is replete with references on the effectiveness of composting in killing off disease-causing agents. However, reports by individuals who actually conducted research on the subject are rather sparse in number. Most of the reports on work on the heavy metals and toxic substances are in the literature on the land disposal of sewage sludge. Although they deal with sludge, the principles involved are applicable to the use of the compost product as well.

An example of the metal concentrations that might be found in compost is taken from a table in a report on work done at the Public Health Service — TVA demonstration plant at Johnson City, Tenn.[62] The table lists elements contained in a 42-day-old compost. The following items, expressed as percent dry weight, were selected as being pertinent here: *Composted mixture of refuse and sludge:* sodium (Na), 0.42; calcium (Ca), 1.41; magnesium (Mg), 1.56; iron (Fe), 1.07; aluminum (Al), 1.19; copper (Cu), less than 0.05; manganese (Mn), less than 0.05; nickel (Ni), less than 0.01; zinc (Zn) less than 0.005; boron (B), less than 0.0005; and mercury (Hg) and lead (Pb), not detected. *Composted refuse only:* Na, 0.41; Ca, 1.91; Mg, 1.92; Fe, 1.10; Al, 1.15; Cu, less than 0.03; Mn, less than 0.05; Ni, less than 0.01; Zn, less than 0.005; B, less than 0.0005; and Hg and Pb, not detected.

Pathogens that might be encountered in improperly composted wastes would be mostly enteric in origin. The reason is quite obvious if the raw wastes were manures, and less so if they were municipal wastes. Naturally, if crop wastes alone constitute the raw wastes, chances are slight that enteric organisms would be present. Theoretically, refuse should be relatively free of enteric microorganisms because practically all state and local regulations forbid the discard of human excreta with refuse. However, one important loophole exists in the form of the discard of disposable diapers. Despite the general misconception that infant feces is in a class by itself in terms of freedom from pathogens, the bacterial population of the stool of an infant older than a few weeks accurately reflects those of the adult members of his or her family.[63,64] Consequently, the hazards inherent in the presence of feces from adults also attend those of feces from infants.

Another and more plentiful source of enteric microflora is the excreta from the huge pet population in the United States. In comparison to this source, the contribution from disposable diapers becomes minute.[64,65] In studies conducted at the University of California which involved making chemical and microbiological analyses of leachates from simulated open dumps and sanitary landfills loaded with freshly milled domestic refuse, concentrations of fecal streptococci in the initial leachate were as much as $1.0 \times 10^9/100$ ml, an amount that compares well with that in raw sewage. This large number, coupled with the similarity of the total coliform to fecal coliform counts, constitutes evidence of the presence of a large amount of animal fecal material.

The relative insignificance of the contribution from disposable diapers was shown by the fact that no difference in enteric bacterial counts could be detected between the leachate from the controls (no disposable diapers) and that from the variables. Some of the variables received 10 times the feces loading to be expected if every infant in the U. S. were supplied with disposable diapers.

Since the chances of adult human feces reaching the solid waste stream are remote, the preceding facts show that the only source of the heavy concentrations of enteric organisms in refuse is of animal origin. The public health implication of this fact is that proper precautions must be followed when composting refuse and using the compost product, even though neither night soil (untreated human feces) nor sewage sludge may have been added to the raw refuse.

Of course, not all disease-causing organisms that may occur in wastes, especially refuse, are of enteric origin. Some may be found in meat scraps (e.g., *Trichinella spiralis*), sanitary paper products (e.g., facial tissues), and improperly disposed medical wastes (e.g., bandaging). Literature on viruses in solid waste and landfill leachates is quite scarce. A major reason for the paucity of publications probably is the difficulty in isolating viruses either from raw refuse or from leachates.[59-62] Nevertheless, viruses of human origin (poliovirus) have been found in refuse.

Types of disease-causing organisms which could occur in wastes (e.g., manure, sewage sludge, and refuse) are especially *Salmonella typhosa, Salmonella* spp., *Shigella* spp., *Brucella* spp. or *suis, Micrococcus pyogenes, Staphylococcus pyogenes, Mycobacterium tuberculosis, Mycobacterium diptheriae, Endomoeba histolytica, Taenia saginata, Trichinella spiralis, Necator americanus,* and *Ascaris lumbricoides.* The last five groups are parasitical forms. Of the genera listed, *Salmonella* and *Shigella* have the greatest significance in composting refuse because they are more apt to be encountered, judging from the heavy concentrations of indicator organisms.

As any devotee of composting is aware, one of the major advantages of composting wastes is the destruction of disease-causing organisms. It is this characteristic that makes composting so attractive for treating certain potentially biologically hazardous wastes (e.g., sewage sludge) that otherwise would be expensive to be rendered pathogenically harmless. The nature and extent of the destruction has been the subject of several studies.[66-71] Judging from the early literature, a characteristic of well-managed composting operations was the absence of health hazards.

Along this line, in the early 1950s the Medical Research Council in England found no health hazards connected with the production of compost. In South Africa, Blair[66] and Van Vuren[72] came to the conclusion that no danger accompanied a well-run compost operation even if night soil were composted. These early studies, however, were largely qualitative in nature, and strong evidence in the minds of the investigators was the absence of fly larvae and the failure of the compost plant workers and users of the product to contract diseases. They had only a few data from bacteriological analyses on which to base their conclusions. Probably the fact that flies traditionally have been associated with the spread of disease constituted the reason for their conclusion.

In studies conducted at the University of California,[73] no attempt was made to isolate pathogens and parasites from the compost piles, although a table was compiled which listed the thermal deathpoints of some of the more common pathogens and parasites (cf. Table 4). On the basis of the information the investigators felt that disease-causing organisms could not survive in a well managed compost pile. An unintended and unforeseen problem that arose from the publication of the table was a subsequent tendency of many writers to offer it as sole evidence of safety of composting without resorting to actual experimentation. This tendency is unfortunate because temperatures in a mass of composting material are not uniform throughout, especially in a windrow. Consequently, even though lethal temperatures may occur in one part, they may not be present in another part. In fact, the indicator organisms (coliforms) may be killed off in significant numbers when the temperatures reach 50°C and higher, only to reappear when the temperatures go down.[43,70]

TABLE 4. THERMAL DEATHPOINTS OF
SOME COMMON PATHOGENS AND PARASITES

Organism	Thermal Deathpoint		Remarks
	Temp. (°C)	Exposure Time in Minutes	
Salmonella typhosa	55–60	30	No growth beyond 46°C
Salmonella spp.	56	60	
	60	15	
Shigella spp.	55	60	
Escherichia coli	55	15–20	
	60		
Micrococcus pyogenes var. aureus	50	10	
Streptococcus pyogenes	54	10	
Mycobacterium tuberculosis var. hominis	66	15–20	
Mycobacterium diptheriae	55	45	
Brucella abortus or suis	61	3	
Endamoeba histolytica (cysts)	55		
Taenia saginata	55–60	5	
Trichinella spiralis	62–65		Infectivity reduced on 1 hour exposure to 50°C
Necator americanus	45	50	
Ascaris lumbricoides (eggs)	60	15–20	

In an investigation on composting sewage sludge at Beltsville, Maryland,[43] it was noted that *Salmonella* which survived in one part of a pile could recontaminate the portion in which they had been killed. However, the investigators did find that repeated turnings led to the eventual destruction of the organisms.

The Beltsville findings that pathogens are destroyed in composting are a confirmation of those made by European researchers a decade or so earlier. Thus, in the late 1950s and in the 1960s, Strauch,[67] Banse et al.,[68] and Knoll,[69] concerned over the disease-spreading potential of composting sewage sludge and refuse, investigated the survival of various species of Salmonella. Banse et al. found that under the conditions of their experiments Salmonella could be rendered harmless within three to five days, and that Bacillus anthracis could survive over a period of seven days. In more recent studies,[71] Mycobacterium tuberculosis organisms were destroyed in windrowed material by the four-teenth day if temperatures of 65°C were reached, and in all cases by the twenty-first day. An exposure of 30 minutes in a pile that had reached 55°C was sufficient to deactivate poliovirus.

In recent studies carried on in mainland China,[74] it was found that composting a mixture of night soil and urban refuse under field conditions resulted in the destruction of Bacillus dysenteria and Salmonella typhosa within a few days after temperatures above 50°C had been reached. While a few ascaris eggs remained viable for one to three days when the piles were at 50 to 55°C, all were killed within a few days when the pile temperatures were within the range 55 to 70°C.

The studies mentioned in the preceding sentences were conducted with the use of windrow composting. The only study available to this author that involved a mechanical digester is one by Shell and Boyd.[75] For test pathogens, they used Salmonella newport, Candida albicans, Ascaris ova, and poliovirus 1. They noted that the test organisms were killed in less than three days.

It cannot be stressed too emphatically that all of these reports are for properly conducted operations. The failure to reach 100 percent pathogen kill is a function of not providing proper conditions. A practical bearing of such a lapse is illustrated by a report written by Peterson.[76] In an examination for parasites in marketed and stockpiled compost, she found nonviable helminth ova and larvae in stockpiled compost. In two out of five marketed samples, she found viable ova of Trichuris trichuris and viable larvae of strongyloides and hookworms. Apparently the samples were of a product that included sludge as well as compost. The lesson to be learned is that a product is not necessarily safe simply because it has been produced in a compost system. A sine qua non is that the system be properly operated.

Although strictly speaking they are not of public health significance, some mention should be made of plant pathogens at this time, inasmuch as they have a very important bearing on the utility of a compost. Moreover, the principles involved in their destruction are the same as those for human and livestock disease-bearing or disease-causing organisms. In addition, since in many cases material to be composted may include residues from disease-infested crops, it would be prudent to know the bearing of these residues on the pathological potential of the product. The information is especially important when it concerns those plant pathogens that permanently contaminate the soil. Examples of such pathogens are *Plasmodiophora brassicae, Alpidium brassicae, Rhizoctonia solani,* and *Heterodera rostochiensis.*

In an extensive study at Diessen, Germany, Martin[77] found that *Plasmodiophora* spores survived at least 253 days if the temperature did not exceed 46.5°C. When the temperature reached 60 to 67°C, no germinating spores could be found after 21 days of exposure. Samples from material that had been processed for seven days in a Dano composter in which temperatures as high as 70°C were reached were devoid of viable spores. These results indicate that a compost that is safe in terms of public health is also safe with respect to crop production.

Heat is not the sole factor in the destruction of pathogenic organisms. Antibiotic reactions and competition between organisms for nutrients are two additional factors. It is not surprising that antibiosis should occur, since composting material usually has an abundant population of actinomycetes and fungi. Competition for nutrients perhaps has an effect that is more inhibitory than lethal. Nevertheless, temperature undoubtedly outweighs antibiosis and competition in terms of significance. The latter two probably accentuate the effects of temperature. Inasmuch as temperatures as high as the 60's and the lower 70's (Celsius) are reached routinely in a compost pile, the thermal deathpoints of all pathogens that have been studied is reached and even surpassed. An element adding to the lethality of temperature is the time factor, since high temperatures not only are reached but also persist for a matter of days, i.e., far beyond the time required to kill the organisms. Nevertheless, Knoll[78] demonstrated that the thermal reaction was by itself not enough to account for all of the destruction of pathogens and that antibiotic phenomena play an important part. In his studies he showed that when the test organisms were placed in a glass vial inserted in a compost pile and thus exposed only to high temperatures, they survived much longer than when placed in direct contact with the other microflora in the pile.

It is the common conviction that composting minimizes fly production. In the University of California studies[16] neither fly larvae nor other evidence of fly development could be found in the composting material. This occurred despite the fact that the incoming refuse was heavily infested with fly larvae. However, in the University's investigation, a hammermill was used that completely destroyed all larvae in the refuse during passage through the machine. Nevertheless, the significance of finding no larvae in the piles is that no further infestation was made by adult flies after grinding. A more convincing evidence for the failure of composting material to support fly production is that of the experience with composting at the Mexican plants. There, despite an abundance of adult flies on the surface of freshly ground windrowed material, no evidence of them in any stage of development could be found within the windrows.

The adverse effect on fly development is largely thermal and on a scale lesser than nutritional in origin. Temperatures reached are much higher than the thermal deathpoint of the fly regardless of stage of its life cycle. As far as nutrients are concerned, the changes in the nature of the raw material brought about by decomposition render it unsuitable as a nutritional source for flies. This explains why, as it often happens, that although adult flies may alight on a pile, they rarely remain long enough to deposit eggs.

An exceedingly important point is that, from the public health standpoint, the compost product is safe for unrestricted use only in proportion to (1) the lack of hazard inherent in the nature of the raw material (e.g., raw sewage sludge as compared to crop residue) and (2) the length of exposure of all the material to the bactericidal conditions prevailing in the interior of a pile or in a digester during the active (high-temperature) stage. If a portion of the hazardous material is not exposed to these conditions, it is likely that it will contaminate the entire product. This warning is especially true if raw sewage sludge is composted with refuse or with any other residue. The practical impossibility of guaranteeing such exposure, short of sterilization by external means, makes it imperative that restrictions be placed on the use of such products. More is said about these conditions in the section on sewage sludge.

SPECIAL APPLICATIONS

In the preceding sections on composting, as well as in the two to follow, the principles and technologies described are for general application. When specifics were discussed in those sections, they usually were concerned with municipal wastes. The present section is a departure from that theme and is concerned with composting four specific wastes: park and garden debris, cannery wastes, dairy and feedlot manure, and human excreta as found in the so-called compost toilets. Although the discussion on the last-named subject would seem to be more appropriate in the section on sewage-sludge treatment, it is included in this section because the methods differ from those for sludge. On the other hand, the composting of sludge is not included in the present section and is discussed in the chapter on the land disposal of sludge. The principles and procedural details described in the previous sections apply to these special applications with the minor exceptions to be described.

Park and Garden Debris

The park debris of concern here consists of small branch and shrub trimmings, grass clippings, and plant residues, but does not include the refuse discarded by the users of park facilities. In other words, it is identical in nature with the garden debris discarded by the home gardener. The scope of the subject includes the leaves collected from the streets. Leaves constitute a major seasonal street-cleaning task for communities located within the temperate zone. Disposal of garden debris is a problem only to the home gardener who neglects composting the wastes or does not have the facilities to do so. Conversely, the disposal of such wastes would not be a problem either to the private or to the public gardener if the wastes were composted. Moreover, the wastes can be composted on a community scale with a minimum of expenditure of capital and effort. Nor is the activity a purely "negative" one, i.e., getting rid of something. It has a positive return in the form of an excellent compost product useful in the most exacting of horticultural applications. Not to be overlooked is the substantial monetary equivalent of the product in terms of reduction of fertilizer and commercial soil-conditioner expenditures. Many communities have long been aware of these advantages and consequently have been composting park debris and leaves for many years.[79-84] In most instances the practice does not have the status of a municipal policy but rather is due to the initiative of the park administrators, and often the city officials are unaware of its existence.

The greater number of the operations deal solely with garden debris generated in public recreational areas and leaves collected from the streets. Only in a few isolated instances does the activity include home garden debris. One of the exceptions is the Westfield, New Jersey, (population 35,000) operation.[79] This community has a center to which its citizens can bring their garden debris to be composted. In return, the citizen receives compost for use in his or her garden. Thus the citizen is benefited in two ways: The need to expend money on disposing the garden wastes is eliminated, and a valuable product is obtained for the garden.

An operation involving a much larger input of material has been recently placed in operation in Berkeley, California, (population 114,000); but the overall procedure is the same: Residents bring their garden wastes to the compost site, to which park debris is also carted. Leaves are not mixed with the debris but rather are stacked separately. The reason is that they include street sweepings which could detract from the utility of the product. Excess compost remaining after the city's needs have been met will be given to the residents in exchange for their garden debris. The operation is made self-supporting through the imposition of a $0.50/cu. yd. dumping fee. The regular fee at a nearby landfill (privately operated) is $0.75 to $1.50/cu. yd.

The equipment and operational needs for composting plant debris and leaves are much less complicated than for municipal refuse because the raw material is easily processed. Moreover, the elemental composition of the raw material is ideal for composting. For example, the C/N ratio rarely exceeds the maximum permissible. (The exception occurs when the amount of dry leaves and wood from chipped tree branches exceeds that of green garden wastes, e.g., grass clippings, plants.) In addition, garden wastes break down readily, and hence special handling is not necessary.

The principal equipment requirements are a grinder and front-end loader. The grinder need not be the massive type required for milling refuse, nor should the milling action of the machine be the same. The ideal mode of size reduction would be shredding rather than the impaction characteristic of a conventional hammermill. Impaction tends to mash green plant material and produce a "soupy" product. The conventional wood chipper is ideal for branches. The only size reduction needed for the smaller garden or park plants is that required for ease of handling. In this author's experience, even dahlia and marigold stalks need be chopped into only 12-in. (30-cm) pieces. A front-end loader instead of a special turner is recommended because the volume of material to be handled would not be large enough to warrant the price of the specialized equipment recommended for composting municipal refuse. Moreover, the front-end loader can be used for other purposes when not involved in turning the windrows. However, if the operation exceeds 40 to 50 tons per day (36 to 45 metric tons/day), the purchase of a specialized turning machine should be considered.

Dairy and Feedlot Wastes

The management of dairy and feedlot wastes can have a disastrous environmental impact unless properly accomplished. If the wastes are improperly stored, they serve as an attractant, abode, and source of nutrient to flies. If the stored wastes are not placed upon a paved lot suitably drained and equipped to collect the drainage, runoff, and leachate from the wastes, they can constitute a significant source of nitrogen in nearby surface and ground waters. In fact, it has been surmised that even the ammonia volatilized from the wastes and wafted about by air currents has been a source of nitrogen pollution in adjacent waters by way of being dissolved in rain falling on the lakes.

The problem of disposal is more acute with operations involving large numbers of animals not only because of the amounts of waste to be handled, but also because of proportionately less land area on which to dispose of them. Generally, the operator of a small dairy, and less so of a small feedlot, has sufficient area on which to dispose of the wastes by land spreading. In dairying, a sizeable fraction of the total manure production is left in the pasture on which the animals graze. In a large-scale dairy operation, however, grazing is at a minimum, and so most of the wastes are concentrated at one site.

A major problem in the disposal of manure is common to all sizes of operations, namely, that of the nature of the wastes. Their consistency is such as to render them readily amenable neither to dry nor to wet handling and hence to methods of treatment associated with each of the two modes of handling. To utilize a system designed to dispose of solid wastes, moisture must be removed from the manure; and for one used in treating wastewaters, water must be added. Either approach becomes costly — with the former it is the cost of the energy to remove water; and with the latter, it is the cost of the added water.

The reference to costs brings up another important problem in animal-waste disposal. It is that of the limitations imposed by economic feasibility upon the cash expenditure for treatment. Economic feasibility in this case is determined by the profit margin the farmer has in his operation, which unfortunately with the present economic setup is rather small.

The aforementioned constraints sharply curtail the technology that can be applied to animal-waste disposal. Under them, the optimum method would be disposal on adjoining land at loading rates within the limits of the capacity of the land to assimilate the wastes without an unfavorable impact on the environment or detriment to crop yield. However, as stated before, this method is available only to small operations. Disposal on land not in proximity of the operation requires some pretreatment of the wastes — usually drying to facilitate handling. Other systems are incineration, pyrolysis, anaerobic digestion, and composting. Incineration and pyrolysis are neither economically nor energetically attractive, and they are wasteful of a valuable resource. The high capital and operating costs make the two methods very sensitive to economies of scale, and the breakpoint in the scale is at a level much greater than that possible with most animal-products operations. Aside from the capital and operational costs is the energy cost to sufficiently dry the manure to permit combustion and effective pyrolysis. Anaerobic digestion probably is satisfactory for large-scale operations in terms of environmental impact and energy costs. More is said about this approach in a later chapter. Finally, there remains composting. Composting is not sensitive to the scale economy which limits the scope of application of the other technologies.

While composting may be said to be a satisfactory method of treating animal wastes, it is not without its difficulties, some of which are sizeable. The most serious problem is that unless well mixed with bedding, fresh manure (cattle, horse, and poultry) is too wet and amorphous to compost as is. The implication is that it must be dried or be mixed with a dry absorbent material (straw, dried leaves, sawdust, or wood chips). Mixing manure, especially cattle manure, with an absorbent material is a task not easily done. However, once the drying or mixing has been accomplished, the compost process becomes a straightforward, uncomplicated operation. The manure is regarded as sufficiently dried when, if handled, it does not compact into a solid mass or form balls or adhere to a shovel. The problem is to maintain the material at a moisture content high enough to permit composting, and yet low enough to minimize compaction. With dairy cattle manure, this level would be about 60 percent moisture.[41] The techniques and principles described in the preceding sections are applicable here. The product is a high-grade, weed-free compost having a very good eye appeal, and hence is readily marketed.

Cannery Wastes

Usually, cannery wastes have a very high moisture content, often to the extent of being in the form of a slurry. Therefore, cannery wastes generally are too wet to be composted alone. For example, peach and apricot wastes have a moisture content of about 85 percent.[17] However, if incorporated with an absorbent material to produce a mixture having a satisfactory C/N ratio, moisture content, and porous structure, the composting proceeds readily. The compost product is of excellent quality both in content and in appearance. The rate of composting can be accelerated by adding lime to counteract acidity and by enriching with urea or other nitrogen source to raise the C/N ratio.

Among the usable absorbents for cannery wastes are municipal refuse, chopped straw, sawdust, rice hulls, and dry leaves. Of course, the amounts of the latter to be added depend upon the moisture of the wastes. The proper moisture content of the mixture is determined as described earlier in this chapter in the section headed "Principles." Naturally, the selection of an absorbent should be decided by the availability of a given waste. Availability involves not only proximity but also economics. Of the absorbents, straw and sawdust probably would be the most expensive. Municipal refuse is always with us and is readily obtained. The availability of leaves is a seasonal occurrence. Of the materials, municipal refuse is the least desirable because of its limited absorbing capacity and mostly because of the expense and effort to prepare it. However, if a complete resource recovery operation is at hand, then it would be worthwhile to consider refuse.

Sawdust, straw, and rice hulls can be utilized repeatedly. This is done by using the composted mixture as the absorbent for new loadings of cannery wastes. Material of the resistance of straw, rice hulls, and sawdust does not break down at the very rapid rate characteristic of cannery wastes. In fact, it takes a large number of passes for such absorbents to reach a state in which they no longer serve their purpose. By the time the absorbent has reached a condition that calls for replacement, the compost has become a top-grade product extremely effective as a soil conditioner, especially for heavy soils.

If lime is added, it is done to serve as a buffer against the drop in pH level, which may be pronounced when fruit wastes decompose. Studies by the National Canners Association researchers indicated a reduction in the initial lag period that occurred with unlimed material.[17] This lag was coincidental with pH levels as low as 3.8 and 4.0. When lime was mixed with the raw material, the pH never dropped below 6.5; and the lag period did not take place.

Nitrogen may be needed to lower the C/N ratio, especially if the compost is to be added to the soil. However, if the fresh absorbent material consists of resistant material, or if the material after being composted is destined to be recycled as an absorbent, then the permissible C/N ratio could be higher than that indicated for composting in general. In the National Canners studies, the C/N ratio of mixtures of peach or apricot and sawdust or rice hulls was as high as 1,000:1 in some cases. The reason is not hard to find: The sawdust had a C/N ratio of 7,900:1. Nitrogen may be added in the form of manure, green vegetable matter, or as a chemical. The National Canners Association workers found that urea served the purpose satisfactorily. If sorted municipal refuse is added as an absorbent, especially that from which the usable paper has been removed, then no nitrogen need be added.

Compost Toilets

A concern about resource conservation has led many to question the practice of degrading the large amounts of water used in supplying the nation's flush toilets. As a consequence, an interest has been aroused in the development of small-scale, excreta management systems that require little or no water. In addition, the need for a toilet that could be used in areas where sewage systems are nonexistent and where septic tanks are not feasible has prompted a similar interest in many not necessarily concerned about the degradation of water. While the old-fashion outdoor privy might seem to be the only recourse for the affected individuals, it has too many objectionable features, which range from public health hazards to less than acceptable aesthetics.

In an attempt to fulfill the need for a substitute for the flush toilet, modifications of the outdoor privy have been made in which composting the excreta is the essential element. Common to the design of the toilet complexes proposed thus far is a sizeable storage chamber or chambers in which the excreta are deposited. In some cases, the storage chamber may be large enough to accommodate a two- or three-year output of a household's excreta. Provision for aerating the mass of excreta is made either by an arrangement of ducts and provision of circulating fresh air or by periodic manual turning. The aeration is intended to promote aerobic composting and the attainment of the attendant benefits. With those providing "passive" aeration, time is designed to become the key factor in stabilizing the excreta and rendering it safe for disposal in the environment. The disposal hopefully (i.e., with respect to the promoter) would be as a soil amendment. The storage period is shorter with those involving active aeration (i.e., true composting), and the stabilized (composted) product is intended to be used in the same manner as any other compost product.

Another feature common to "compost" toilets is the requirement for material in addition to excreta to serve as an absorbent and hopefully to provide some structural strength to the mass. In one system, this is provided by layering the bottom of the chamber with the absorbent at the time the unit is first placed in operation. Kitchen wastes provide a continuing source thereafter. This is made possible by the installation of a second port through which the kitchen wastes may be dropped. A system that calls for manual turning prescribes the addition of an absorbent (e.g., sawdust) with each use. As the user prepares to make use of the facility, he brings in a handful or so of sawdust to be thrown into the hole.

Two of the better known systems are the Clivus and the Van der Ryn. The Clivus system[85] basically consists of a large compost tank having a sloped bottom and equipped with incoming toilet and kitchen wastes tubes, an outgoing air vent, and an unloading door. The main tank is divided into three interconnecting chambers. The first receives toilet wastes; the second, the kitchen wastes; and the third provides a storage chamber for the combined wastes. When the unit is in operation, the toilet wastes slowly slide downhill into the kitchen wastes; and the two combine and further decompose as they slide into the storage chamber. The unit is designed to have a volume such that it is two or three years before stabilized wastes appear in the storage chamber. An average family of four supposedly could use the unit up to 10 years before having to remove the stabilized product.

Although the manufacturers of the unit claim that aerobic composting takes place in the chambers, the nature of the fecal-kitchen waste mixture and the arrangement and construction of the aeration ducts are such that the greater mass of the wastes inevitably remains anaerobic. The most that can be said is that the chambers are well ventilated. The excellent ventilation effectively prevents the travel of malodors into the bathroom or kitchen. The term *composting* here can be applied only in the broad sense of the term, *decomposing* is a more appropriate term. As far as public health is concerned, the saving feature with respect to the use of the "compost" product is the time factor. However, should the unit be loaded such that the retention time is materially shortened, or if short-circuiting should occur, then the "compost" product should be handled with precautions identical to those with night soil.

The Van der Ryn system incorporates composting in the strict sense of the term and, if properly operated, is characterized by aerobic composting. The superstructure is much like the conventional outdoor privy. The infrastructure consists of two chambers. One receives the excreta and the second serves as a storage chamber. A container of sawdust or other absorbent material is kept next to the toilet bowl or "squat plate." Each time an individual defecates, he or she drops some of the absorbent into the collection chamber. Thus a mixture of excreta and absorbent gradually accumulates. The chambers are equipped with removal slats arranged such that very easy access is had to the material for periodic turning.

The disadvantage with the Van der Ryn system is that it relies upon a considerable amount of dedication on the part of the householder, since one would hardly classify the turning as a pleasant chore! However, if the turning is done carefully and faithfully, the overall operation should be satisfactory. Nevertheless, the precautions named in the section on public health should be very carefully observed.

ECONOMICS AND STATUS OF COMPOSTING AS A WASTE-TREATMENT PROCESS

Municipal Refuse

It is difficult to obtain sound numbers on the dollar cost of composting municipal refuse, and the difficulty is aggravated by the inflation factor. The problem arises from the nature of the cost analyses available in the literature. Generally, they tend either to be unrealistically low or inaccurately high. Not unexpectedly, the optimistic analyses are reported by promoters of specific systems or by too-ardent enthusiasts. The excessively conservative estimates are made by individuals who either can see no future for composting under any circumstance or who mistakenly base their calculations on a literal interpretation of a demonstration-scale operation, i.e., do not make allowances for economies of scale. Naturally, the correct answer lies somewhere between the two extremes.

The problem is further complicated by the fact that of the few analyses reported, most are out-of-date. Other reasons for the difficulty are differences in size of plants; in operational methods; in size of the work force, wage scales, and number of work shifts; in methods of accounting; in land costs; and in the method of disposing the product. Probably the most objective analysis of the economics of composting at the time of this writing is to be found in the EPA report *Composting of Municipal Solid Wastes in the United States.*[62] Yet even in this report, many of the numbers are uncertain because of the reasons given in the preceding discussion.

A major consideration in estimating the cost of composting refuse is whether or not composting is to be an integral part of an overall resource recovery operation or whether it is to be the only resource recovery activity. This consideration is important because of its effect on the dollar cost. If it is to be a part of an overall resource recovery undertaking involving grinding and classification, then the cost of the preparatory steps (grinding and sorting) should be apportioned among the subsystems, e.g., ferrous metal, nonferrous metals, glass, paper salvage, and organic material recovery by composting, inasmuch as all of the subsystems share in the need for the preparatory steps. In terms of cost per ton of compost, the reduction would be substantial. At a milling cost of $3.00/ton ($3.35/metric ton), the assessment for composting would be only about $0.60/ton ($0.68/metric ton) if the five groups of resources were recovered, as against the full $3.00/ton if composting were the only aim. In a resource recovery system such as that developed by Trezek and his associates[39] the apportioned cost should even be lower because with their system, only final organic rejects are composted. In other words, only that material is composted which otherwise would have to be landfilled. On the other hand, if only ferrous materials were recovered, then practically the full cost of pretreatment should be charged to the compost component. The reason is that a minimum of processing is needed to prepare incoming wastes for exposure to the magnet with which the ferrous materials are removed. The lesson to be learned from these considerations is obvious: If composting is to be practiced, the best course is to set up a complete resource recovery operation.

■ **Investment Costs:** The EPA report[62] is the source for the numbers given in this section. The costs as given attribute the entire expense for materials preparation to that for composting and, hence, represents the maximum costs. Moreover, they do not represent any credit for sales of the compost product. However, the market for compost heretofore has been practically nil. The estimated costs for a 150-ton/day (135 metric tons/day) plant, excluding land costs and including interest at 7.5 percent for 20 years would be about $1,237/ton ($1,390/metric ton) capacity or $0.48/ton ($0.54/metric ton) refuse processed with a windrow system as operated at the United States P.H.S.-TVA Johnson City plant; and $2,005/ton ($2,230/metric ton) capacity of $0.75/ton ($0.83/metric ton) for an enclosed (mechanical) operation of the type at Gainesville, Florida, (Metro Waste Conversion System).[86] If sewage sludge is processed with refuse, then the estimated cost per ton (1969 dollars) with a windrow system would range from $5.38/ton ($5.99/metric ton) refuse processed in a 50-ton/day plant (45 metric tons/day) to $3.10/ton ($3.45/metric ton) in a 200-ton/day (180 metric tons/day), one-shift plant or $1.73/ton ($1.92/metric ton) in a two-shift plant. Cost per ton daily capacity for the respective plants ranged from $15,560 down to $4,600. Again it should be stated that these costs do not necessarily apply to other situations, especially with other types of enclosed systems. For example, with some enclosed systems, the capital cost may be as high as $10,000 to $15,000/ton ($11,100 to $15,866/metric ton) daily capacity for a 150-ton/day (135 metric tons/day) plant. Nevertheless, the numbers do give an approximation of the costs to be expected.

■ **Operating Costs:** Representative costs for operation are even more difficult to ascertain than is the case with capital costs. Here, labor costs assume an important part. For example, in the Johnson City U.S.E.P.A.-TVA project it accounted for 75 to 78 percent of the operating costs.[62] However, at the time the Johnson City plant was in operation, turning machines of the capacity of the Terex-74-51 were not available. The use of such a machine should reduce the labor cost from 25 to 30 percent of that required in the Johnson City operation. According to estimates in its EPA report,[62] operating and maintenance costs for a windrow plant following the procedures used at Johnson City would range from about $10.00/ton ($11.11/metric ton) (1969 dollars) with a 100-ton/day plant to $5.00/ton ($5.55/metric ton) with a 300-ton/day plant. With a Metro-type enclosed system as operated at Gainesville, the cost would range from $7.56/ton ($8.40/metric ton) at 157 tons/day (141 metric tons/day) to about $6.94/ton ($7.72/metric ton) at 346 tons/day (320 metric tons/day).

More recent (September 1973) costs are available that pertain to a sewage-sludge compost operation in which sludge is composted with wood chips.[43] Operating costs for a 600-ton/day (540 metric tons/day) output, including the cost of the wood chips but excluding that for transportation of sludge, are $350,000/year and $1.65/ton ($1.85/metric ton). This includes labor, fuel, oil, compost analysis, and electricity costs. Although these figures are for composting sewage sludge, they furnish an indication of what the costs would be for composting municipal refuse as a part of a total resource recovery plant. The comparison is justified because it reflects the major item of cost in such an operation, namely, turning the windrows.

■ **Market:** In the 1950s, promoters of compost systems offered the possibility of profits from the sale of the compost product as a lure to prospective municipal customers. Instead of being an expense to the municipality, refuse disposal would be a source of income, provided the wastes were composted. This motivation proved in the long run to be a great obstacle to the adoption of composting, because the promised profit failed to materialize. The reason was an almost complete lack of a market for composted municipal refuse. Although no market exists for composted refuse, there is a significant one among home gardeners, orchard growers, and truck farmers for properly composted manure and other nonrefuse organic wastes.

One of the many reasons for the lack of a market for composted refuse is the generally poor quality of the product. The inferior quality is partly due to the composition of the refuse, especially from cities in industrialized countries. Bits of glass, but mostly the pieces of plastic sprinkled throughout the product, detract from its appearance. If a large part of the paper fraction is composted with the other organic matter, the finished product assumes a fluffy, fibrous character unlike that of composted manure or garden debris. On the other hand, composted refuse from a city in a less industrial country such as Mexico has a pleasing appearance, especially if it has been well screened. The reason is that the proportion of food wastes and garden debris to paper is much greater than in the U.S. The quality of composted U.S. refuse can be materially improved by mixing sewage sludge, manure, or other organic waste rich in nitrogen with the refuse. Yet another way is to use the resource recovery system developed by Trezek and his associates[39] in which only the organic rejects are composted. With their method, glass and plastic contaminants are at a minimum.

The inferior quality of the composted refuse is not the only factor accountable for the lack of a cash market. A major reason is traceable to large-scale agricultural practices in the United States. The entrepreneur of a large-scale farm is not interested in bulky fertilizers nor in soil amendments, and compost fits both categories. Although the situation probably will change somewhat as the supply of chemical nitrogen dwindles and consequently its cost increases, it is doubtful that a demand would exist that would be large enough to absorb the entire output of municipal compost if all cities were to switch to composting. The collective demand from all home gardeners and orchardists, if one were developed, probably could be met by the output from one or two fair-sized municipalities.

On the other hand, there undoubtedly is a good market for composts other than those produced from refuse. Composts already on the market are expensive and often leave much to be desired in way of quality, although they do have excellent eye appeal. Consequently, with the development of a good sales organization, no problem should be encountered in selling composted park and garden debris and manures.

The discussion on marketing leads to the next logical subject, and that is the status of composting. As with marketing, in speaking of status a distinction should be made between municipal-scale composting practice and other types of composting; or in terms of wastes, between composting refuse and composting other wastes. Although no firm figures can be given, mainly because none are reported in the literature, it can be safely stated that a significant fraction of steer manure is being composted, even though some of the "compost" product on the market scarcely merits the term *compost*. A survey of the number of home composters undoubtedly would reveal an impressively large figure, probably on the order of a million.

Unfortunately, the status of composting as a municipal practice in the United States is one of nonexistence. At the time of this writing, no municipal-scale plant was in operation in the United States. The picture is less grim on a world-wide scale, although the pattern has been that the more industrialized a nation becomes, the less is the amount of its municipal wastes that are composted. Mexico is an exception in that despite its growing industrialization, it has embarked upon composting on an appreciable scale. The future is brighter there because an attempt is made to integrate agricultural use of the product with compost production. The approach is to foster cooperation between farmers and municipalities.

The major reason for the poor showing of municipal composting in the United States today is the unfavorable economics described in the preceding paragraphs. A contributing factor is what is in effect a double standard as applied to composting. Other systems of solid waste treatment and disposal are expected to cost money; whereas a compost operation is expected to earn money or at least to break even. If it does neither, it is branded as being a failure. The fact that a resource is recovered and a useful product formed receives little or no consideration. If it is any consolation to the compost enthusiasts, there remains the fact that although no municipal-scale plant is in operation in the United States, a significant fraction of the nation's organic wastes (exclusive of paper) is finding its way to the compost piles on the premises of homeowners.

Applications Other Than Municipal Refuse Disposal

The economics of home composting have little bearing on whether or not the average householder does any composting. The major expenditure for the householder is "muscle power," and the deciding factor is the individual's ambition. Expanded to the large scale of composting park debris, economics still does not play a major role, although excessive cost might rule it out. An excessive cost might come from the need to purchase an expensive piece of equipment not already available to the park department. Such a piece of equipment might be a shredder for use in programs such as the one at Berkeley. However, the grinder need not have the stamina of one used in size-reducing municipal refuse. A machine designed for agricultural or institutional applications would be suitable. Tree limbs and saplings already are being chipped by city-owned chippers in most communities, and hence there would be no need to purchase one for a compost project. If the influx of raw material is intermittent and storage space is available, then it might be advisable to rent the grinding equipment as the need arises. Perhaps the major expenditure would be in manpower. Even this expenditure would be moderate and would be limited to the time spent by the operator of the front-end loader while turning the compost pile, i.e., on the order of 10 to 20 percent of his time for a 50-ton/week (52-week basis) operation.

It goes without saying that the economics of composting animal wastes depends upon the nature of the waste to be composted and the size of the operation. Other factors pertain to land use, namely, the amount of available land area and its monetary value. If land is expensive, then the trade-off must be determined between the cost of speeding up the compost process and that of the land released or not needed as a result of the acceleration.

The nature of the material is important because it determines the amount of preparation needed to ready the waste for the compost operation. For example, if feedlot manure is to be composted, either it must be dewatered or it must be mixed with an absorbent material. Here, it should again be pointed out that mixing such manure as it comes directly from the lot with an absorbent is not an easy task. The task is lightened considerably if the material is dewatered to 40 to 60 percent. According to Scholz,[87] approximately 5.16 liters of water must be evaporated from feedlot wastes to produce 1 kg of "humus" (dry) manure; from hen wastes, 2.20 liters; and from swine wastes, 8.0 liters. The reduction of such a moisture content to the level of 40 to 60 percent would require the removal of about 4 liters of water/kg dry weight. To remove all of this water, that is to reduce the moisture content to zero percent, would require an energy input equal to about 4,600 Btu/lb. Obviously, bringing the moisture content to a level intermediate to the existing level and zero percent would require a correspondingly lesser energy input.

The monetary cost of the energy would vary with that of the energy source. According to one report,[88] the cost of heat-drying in southern California varies from $12 to $25 per ton ($13 to $28/metric ton). The energy costs can be reduced considerably through the use of composted manure as an absorbent, and thereby using the compost process itself as the dewatering system. With a Terex-Cobey machine to turn the windrows, the moisture content of a manure can be brought from 80 percent down to 10 to 12 percent in dry weather at a cost of about $0.50/cu. yd. ($0.65/m³) (1973 dollars).

Using the bin forced-aeration system developed by Senn, the total cost (1971) of producing composted manure in his operation amounted to $3.47/cu. yd. (4.54/m³).[41] Packing the material into sacks (13.4/cu. yd. or 17.6/m³) costs $2.97/cu. yd. ($3.88/m³). The total cost of the sacked product was therefore $6.44/cu. yd. ($8.42/m³) or $0.47/sack. Using forced aeration by way of pipes installed in open slabs (windrow composting) brought the cost down to $0.33/sack. Since sacking costs remained constant, the savings came from lowering the composting cost, which was $1.43/cu. yd. ($1.86/m^3). Presently (1975), steer manure in California is retailing at $0.75 to $1.50 per sack.

2. LAND DISPOSAL OF PRIMARY AND SECONDARY DOMESTIC SEWAGE SLUDGES

INTRODUCTION

This chapter is concerned only with domestic primary and secondary sewage sludges to the exclusion of industrial sludges and of tertiary sludges from advanced wastewater treatment systems. The principal reason for the exclusion is that even though the existing treatments for the latter two groups of sludges are not exclusively nonbiological in nature, they are largely so. Furthermore, treatment by anaerobic digestion and disposal by incineration also are not discussed. The latter because it is nonbiological in nature; anaerobic digestion is covered more adequately in standard sanitary engineering textbooks[89-91] than could be done in this text.

The sludges produced in the various steps in wastewater treatment are the solids that settle out of suspension in the course of the treatment. In effect, therefore, they represent a concentration of the settleable solids in the incoming wastewater in its successive steps in the purification process. The significance of this fact is that the concentration includes that of disease-causing organisms, inasmuch as they adhere to or are entrapped by the settling solid particles, and are therefore accumulated in the sludge. The public health significance is thus quite apparent.

Sludges generally are classified according to their source. The various sources are indicated in Figure 13 in which is presented a flow diagram of a typical wastewater treatment system. As the figure shows, there are three major sources of sludge in primary and secondary treatment. The first source is the primary clarifier, and its product usually is termed "raw sludge" (primary sludge).

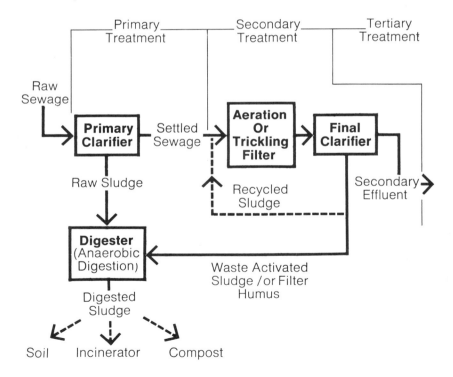

Figure 13: Flow diagram of a conventional wastewater treatment plant

The second source is the sludge from the digester. This sludge commonly is known as "primary digested sewage sludge," or simply as "digested sewage sludge." The third source of sludge is the final clarifier through which has been passed the effluent from the aeration chamber of an activated sludge system or from a trickling filter (not shown in the diagram). If an aeration chamber (activated sludge unit or extended aeration unit) precedes the clarifier, the sludge is designated "waste activated sludge." If aeration is by means of a trickling filter, its sludge is termed "filter humus."

CHARACTERISTICS

Before launching into the presentation on sludges, it should be pointed out that the characteristics of sludges vary widely. This is due to the great differences in types of wastewaters and in the design and operation of wastewater treatment plants.

Raw primary sludge is thoroughly objectionable both in odor and in appearance, and it is biologically very unstable. It has a high moisture content, the extent of which depends partly upon the detention time in the clarifier. Usually, it is within the range of 90 to 95 percent of the wet weight. The waste not only has a high moisture content, it also is very difficult to dewater. Its nitrogen content ranges from 2.4 percent to about 5 percent; and its phosphorus (P_2O_5) content, 1 to 2 percent.[91,92] Its potassium content is negligible.

The physical condition of anaerobic digested sludge is far less offensive than that of raw sludge because digested sludge represents a much higher degree of biological stabilization. Generally, digested sludge as it comes from the digester is dark gray to black, has an odor suggestive of tar, is readily dewatered, and has solids that are somewhat granular in texture. In fact, ease of dewatering is a measure of quality of the digestion.

Solids concentrations of sludge direct from the digester range from 3 to 5 percent. Special applications may result in a concentration as high as 10 percent. Finding and interpreting reports on the chemical composition of digested sludge as well as of the other sludges is a difficult and perhaps frustrating task. The principal problem is in ascertaining the basis on which the values are reported, e.g., on wet weight or on dry weight. In some cases if the reported values are based on wet weight no mention is made of the solids concentration of the sludge. The values given in this text are on a dry basis and are for products that may include digested raw (primary) sludge alone or a mixture of digested raw and waste activated sludge or trickling filter humus. Reported values for nitrogen content range from 2.0 to 5.1 percent; P_2O_5 concentration, from 1.2 to 6 percent; and K_2O, from 0.2 to 1.1 percent.[93,94] Other elements of significance and their concentrations are: Cr, 26 to 1,500 ppm; Hg, trace amounts to 1 ppm; Pb, 16 to 3,407 ppm; Zn, 80 to 6,500 ppm; Cu, 600 to 1,700 ppm; Cd, 13 to 300 ppm.[93-95] The wide range in concentration of metals is a function of the extent to which industrial wastewaters are treated along with residential wastewaters. The more the industrial sewage, the greater is the concentration of metals in the sludge. The higher concentrations of Cu, Zn, and Pb come from the piping used in plumbing. Inasmuch as copper tubing has become the predominant choice in modern building construction, the copper content of wastewaters, and hence of sludges, will continue to be high.

The biological population of digested sludge includes pathogens, but at a far lower concentration than in raw sewage sludge, especially if the digestion process is carried on at an elevated temperature, i.e., 35°C or above. If digestion is carried on at thermophilic levels, the pathogen population is reduced to an insignificant size. At temperatures lower than 30°C, time spent in the digester compensates to a considerable extent for the lack of thermophilic temperatures. Nevertheless, a significant number of pathogens do survive, i.e., enough to warrant caution in the use of the digested product. Since the pathogen populations are so dependent on operational conditions, little is to be gained from presenting an extended list of numbers, and an example or two should suffice. According to one reported analysis, the average total coliform count of the sludge output from a particular treatment plant was 23 × 10^9 per 100-gram sample; of fecal coliforms, 2.4 × 10^9 per 100 grams; and of salmonellae, 6,000 per

100 grams.[43] Although coliforms are not pathogens, they do indicate the potential presence of enteric pathogens, and the chances of pathogens being present are related to the size of the coliform count.

Activated sludges and trickling filter humus generally have a low solids content, and processing them is characterized by many of the difficulties encountered in handling raw sludges. Consequently, activated sludge and trickling filter humus commonly are stabilized by subjection with raw sludge to anaerobic digestion. The fertilizer content of activated sludge and trickling filter humus is on the order of 2.9 percent to 8 percent nitrogen, 3 to 7 percent P_2O_5, and trace amounts (0.3 to 0.6 percent) of K_2O.[91,96] While the concentrations are much less, the variety of pathogens found in the two sludges is comparable to that in raw sludge. This is a result of the fact that the treatment taking place in the two processes has little of the bactericidal about it. Consequently, in terms of handling and utilization, the two sludges should be handled with the precautions followed with raw sludge.

■ **Tertiary Sludge:** As the name implies, tertiary sludges are those produced in the course of tertiary treatment of wastewaters. A description of the various types of tertiary treatment may be found in sanitary engineering textbooks.[89-91] As far as this text is concerned, suffice it to state that the sludges have a high concentration of chemicals, usually of aluminum complexes, or of calcium compounds, and perhaps of iron. The presence of aluminum, calcium, and/or iron is due to the fact that they are the active constituents of the reagents (lime, alum, and ferric chloride) used to bring about the precipitation which constitutes an important element in tertiary treatment. A substantial portion of the calcium is recovered by calcining lime sludges. With the precautions described for the activated sludges, the lime sludges could be used in those agricultural applications in which lime is needed. Recovery of aluminum from sludges is more difficult and hence is not widely practiced. Other than the hygienic aspects, the effect of using these sludges on the soil is the effect on soil structure caused by an excess of the two elements.

DISPOSAL

Sludges may be disposed by wet oxidation, incineration, landfill, and disposal on the land surface either directly or preceded by composting. Wet oxidation as a sludge disposal process is limited to a few isolated operations and to special applications. At present, incineration is the most common course followed at the larger treatment plants, usually because no land is available for disposal. The problem with incineration is that a combination of expensive energy sources and the need for rigid air pollution control is making it a prohibitively expensive process. Consequently, communities are searching for less-expensive alternatives. One alternative is to bury the material in a sanitary landfill. The problem here is lack of a site which can fit within the sociological, hygienic, and economic constraints peculiar to such an operation.[97] Because of these difficulties, the alternative of spreading on the land begins to appear more attractive, even if it means transporting the sludge over appreciable distances, as is being done at present by the Metropolitan Sanitary District of Greater Chicago.[98,99] Land spreading may be done directly after the sludge is discharged from the treatment plant or may be preceded by composting.

The discussions which follow are based on the use of digested sludge, except where specifically noted. Because of the inherent hygienic hazards and for aesthetic considerations, raw or activated sludge should not be applied directly to the soil. The dangers are lessened if the sludge is composted.

LAND SPREADING

Methods

The methods described herein involve sludge in its liquid form. In the dry form it is applied much as are manures. Often the site of the application of the sludge is not in the immediate vicinity of the sewage treatment plant, and hence it must be transported to a distant site, perhaps 100 or so miles away.[98] For nearby applications the material can be piped to the site. Pipe, truck, and barge can be used for distant sites. If past experience with attempts to set up transport systems for municipal refuse are indicative, the use of rail transport would be too expensive. In the Chicago study,[98] truck and barge transport were used. No instances of long-distance transport of sludge by pipeline have been reported in the literature at the time of this writing.

Various methods have been tested for applying the sludge to the soil.[100,101] Among the methods are the following: (1) the use of a tank-truck sprinkler equipped with a manifold across the rear, (2) one or more of a collection of various irrigation systems, and (3) a system for incorporating the sludge directly into the soil.

The sprinkler-truck method is hampered by the need to drive the truck up and down the fields. This compacts the soil and is limited to dry periods when the truck would not be mired in the soil. Simple irrigation by flooding is characterized by a tendency of the sludge particles to settle out of suspension at the point of discharge. This results in an uneven application of solids across the field. Although a ridge and furrow system in which the sludge is directed into the furrows works quite well, it does involve a considerable amount of land preparation. Systems involving the use of sprinklers are plagued by the tendency of sludge to plug the orifices of the sprinkler. Lynam[100] found that nozzles that are 1.5 to 1.9 in. (3.8 to 5 cm) in diameter and are operated at nozzle pressures of 50 to 80 psi (345 to 552 kN/m^3) will not plug.

A soil incorporation system employed in the Chicago operation involved a manifold mounted on a heavy disc harrow or two-way plow. The manifold delivered the liquid sludge to the inside of each blade or mold board which then covered the sludge as the unit moved forward. Other types for incorporation into the soil depend upon the use of a ribbon augur.[101] The advantages of direct incorporation are: (1) the danger of ponding is lessened. (2) The material is out of sight and contact with the external environment, thus lessening the likelihood of aesthetic nuisances. (3) Soil microbial processes can begin immediately to make the nutrients available to the crop and bring about those reactions that favorably affect the structure of the soil.

■ **Loading:** Amount of loading may be considered from two viewpoints: (1) that which is optimum for crop production and (2) use of the soil simply as a disposal medium. The first consideration is based on utility to the crop, and the second on the maximum assimilative capacity of the soil.

The frequency of application is determined by climatic conditions, crop planting programs, and permissible amounts of loading. The constraints imposed by climatic conditions and crop program are obvious. Those from permissible loading require further elaboration.

An important factor in determining amount of sludge per application is the capacity of the soil to absorb the slurry without becoming flooded, i.e., without ponding. The more sandy the soil, the more liquid it can absorb. Of course the danger here is that of leaching to the ground water. Clays accommodate less of a load per application.

A second factor stems from the amount of nitrogen and other plant nutrients that can be applied without leaving an excess to be leached to the ground water. Some disagreement exists as to the amount of sludge that can be applied without leaving mobile nitrogen to be carried to the ground water. Setting up a hypothetical model to predict the loading does not always lead to an answer that fits practice. For example, one such model was developed which specified that only a little more than 0.5 in. (1.25 cm) of sludge per year could be added without leaving some nitrogen to be carried in the leachate.[102] On the other hand, actual practice has shown that about 10 to 15 tons of sludge/ acre/year (22.4 to 33.4 metric tons/ha/year) might be needed to meet the needs of a nonleguminous crop,[103] and thus leave no nitrogen for the leachate. This amount would be equal to 2 to 3 in. (5 to 7.5 cm). In fact, in another study up to 160 tons/acre (358 metric tons/ha) were mixed with soils having a clay subsoil without resulting in nitrogen leaching to the ground water.[103]

The conclusion to be drawn from these examples is that no hard and fast number can be given for loading for a specific application, because the permissible loading is a function of: (1) the nitrogen content of the sludge, (2) the nitrogen needs of the crop (varies from crop to crop) and climate, (3) the nitrogen already in the soil, and (4) the nature of the soil and its substrate. For example, corn may take from 150 to 250 lbs./acre (168 kg to 280 kg/ha). Under "normal" conditions an average value would be about 20 tons/acre-year (44.5 metric tons/ha) since that amount could be assimilated by the soil.[104]

A more serious limitation on total amount, if not on rate of loading, is the possibility of a buildup of metals in the soil. The data in Tables 5 and 6 indicate the type and amount of metals that may be added to the soil through the disposal of sludge on land.

The permissible loading varies according to the metal content of the sludge, the soil type, the pH level of the soil, and the type of crop to be planted. When considering the metal loading to the soil, it should be remembered that a variety of significant sources can be named: (1) airborne emissions from factories, automobiles, and smelters; (2) solid and liquid waste discharges from industries and municipalities; and (3) natural runoff, which may carry pesticides, fertilizers, mine wastes, fly ash, and animal manure. The content of the sludge depends upon the sources of the wastewaters treated at the sewage treatment plant. Thus the sludge from treatment plants serving industrial cities generally have a high metal content. The range of the metal content is indicated by the data listed in Table 5.[105] The amounts of a typical sludge that could be added to a soil without enriching it with given trace elements beyond the concentration normally present in soils are presented in Tables 6 and 7.[106]

At the time of this writing, definite conclusions on the possibility of long-term toxic effects of metals on crops and on animals and humans who consume the crops had not been reached. However, in a discussion on metal toxicity, it should be kept in mind that all living beings require a variety of metals in trace amounts for their continued survival; but when amounts in excess of traces are taken in, those metals become toxic. British agricultural agencies using zinc toxicity as a base reference point, recommend that the total of copper, nickel, and zinc added be such as not to bring the concentration above the toxic equivalent of 500 lbs. of zinc/acre (560 kg/ha). Some investigators doubt that the careful use of sewage sludge on crop land would result in toxicity in the soil. A study in England[107] showed that despite the

TABLE 5. COMPOSITION OF ANAEROBICALLY DIGESTED SEWAGE SLUDGE[95,105]

Element	Calumet Concentration (ppm) (dry weight)	Stickney Concentration (ppm) (dry weight)
Cd	3.0	14
Mn	8.0	18
Ni	3.0	15
Zn	83.0	223
Cu	16.0	67
Cn	26.0	194
Fe	726.0	2100
Pb	16.0	75
Hg	0.063	0.275
Na	98.0	131
P	757.0	1,141
Ca	963.0	1,289
Mg	180.0	484
K	195.0	390
Total N	1500.0	1560 (in solids, 3–4%)
NH_4^t-N	650.0	500–2000
Total C	6,400.0	——— (in solids, 22–27%)
Solids	20,500?	43,600
Volume	580,000?	480,000

application of a total of 568 tons (511 metric tons) of solids from 1942 to 1961, very little effect was noted on the crops, although there was some increase in metal uptake by the plants.

Another aspect making it difficult to assign specific loadings is the effect of the solubility factor on the metal uptake by plants. Plants can assimilate metals only when the latter are in solution. (For that matter, to reach the ground water, the metals must be in solution.) This state of things, coupled with the fact that plant species and even varieties within the species differ in their uptake of metals, further complicates the assignment task. The task is aggravated by a lack of practical information in that studies on the uptake of metals by plants have been mostly on a greenhouse scale, whereas long-term field-scale studies are needed to clarify the situation. Fortunately, the situation is beginning to change and

TABLE 6. TRACE ELEMENTS ADDED TO SOIL TO A DEPTH
OF 15 CM FROM 100 METRIC TONS OF TYPICAL DOMESTIC
SEWAGE SLUDGE WITH AMOUNTS COMMONLY PRESENT[106]

Element	Concentration in Sludge µg/gram	Amount Applied to Soil kg/ha	Amount Present in Soil (kg/ha)	
			Normal Range	Typical Level
Ag	10	1	0.02–10	0.2
As	5	0.5	0.2–80	12
B	50	5	4–200	20
Ba	1000	100	200–6000	1000
Cd	10	1	0.2–1.4	0.12
Co	10	1	2–80	16
Cr	200	20	10–6000	200
Cu	500	50	4–200	40
Hg	5	0.5	0.02–0.6	0.06
Mn	500	50	200–8000	1700
Mo	5	0.5	0.4–10	4
Ni	50	5	20–2000	80
Pb	500	50	4–400	20
Se	1	0.1	0.02–4	0.4
Sn	100	10	4–400	20
V	50	5	40–1000	200

TABLE 7. AMOUNTS OF TYPICAL DOMESTIC SEWAGE NEEDED
TO BE ADDED TO SOIL TO EXCEED THOSE CONCENTRATIONS
NORMALLY PRESENT IN SOILS[106]

Element	Amount to Exceed Normal Concentration in Soil (metric tons/ha)
Cd	140
Zn	300
Cu	400
Hg	120
Pb	800
Ni	1600
Cr	1000
Se	400
Mo	800
B	400
Ba, Co, As, Mn	1000

[1] Unrealistically high loadings required to enrich soils beyond the normal level

at the time of this writing several excellent articles have appeared that deal with field tests.[108-112] Such field-scale studies are needed to clarify the situation.

An important factor with respect to solubility of a metal compound is the pH of the soil. At low pH levels, i.e., in acid soils (pH lower than 5.5), the metal salts are in a soluble form. As such they not only are available for assimilation by plants, they also can be leached to the ground water. (The actual amount carried in the leachate further depends upon the ion exchange and adsorption capacity of the soil particles.) The availability results in an inhibitory response of the crops to much lower concentrations of metals than is the case when the pH level is above 7.0. At pH levels above 7.0, most of the metals and especially the heavy metals are complexed into insoluble compounds, and as such they are far less available to plants.

Generally, well-tilled (aerated) soils rich in organic matter have a high pH level and can bind a sizeable concentration of heavy metals. Soils also can be rendered alkaline by the addition of lime. An extreme effect of liming is that of the uptake of zinc by Swiss chard.[107] At an application of 80 tons sludge/acre (178 metric tons/ha), the uptake of zinc without liming was 1,060 ppm; and with lime, 184 ppm. At 160 tons/acre (356 metric tons/ha), the levels were 1,690 and 627 ppm respectively. Some metal uptake is possible at pH levels in the lower alkaline range because of a shallow, somewhat acidic zone at the surface of the root hairs. The acidity results from the release of respiratory CO_2, which combines with the soil moisture to form carbonic acid.

Except in sandy or very acid soils, the movement of metals added in the form of sludge is restricted almost entirely to the tillage zone. The mobility of Mo, As, and Se, although normally rather limited, may be greater than that of cationic trace elements, especially in neutral and alkaline soils.[106] Boron is quite mobile in most soils. The danger of trace elements reaching ground water is greatest where shallow water tables occur beneath sandy soils. Contamination of surface waters could come from runoff from sludge-amended soils, since the trace elements are concentrated in the surface layer of the soil.

■ **Plant Uptake:**   A sufficiently high concentration of certain of the elements, e.g., Mo, Se, and Cd, can be assimilated by crops to render them toxic to man and animals eating them. Others are assimilated by plants to a lesser extent. For example, in one study involving corn grown on sludge-treated soil and on a control plant, the concentration factor for Mn was 3.7; Zn, 3.5; and Cu, 1.7. The cumulative factors for Mn and Zn generally are greater than for Cu, Pb, Mn, and Cr. The literature contains a multitude of additional examples, most of which are summarized by Page in his review.[106] Usually, the leafy tops of plants accumulate a heavier concentration of a given metal than do the roots or fruits. Thus, Hinesly *et al.*[113] report that at sludge loadings equivalent to the addition of 132 kg Mn/ha, concentrations of Mn in leaves of corn plants were as much as 116 micrograms/gram, but only 18 micrograms/gram could be found in the grain. The amounts of Zn were 212 and 152 micrograms/kg respectively; of Cd, 11.5 and 1.0 micrograms/gram; and of B, 44 and 6.6 micrograms/gram. Using soybeans as the test crop and applying Cd to the soil at 129 micrograms/gram, Jones *et al.*[114] found that the soybean leaves contained 18.5 micrograms Cd per gram of dry matter, and the seeds only 1.0 microgram/gram.

In summary, it can be stated that (1) on the basis of accumulated evidence to date, the uptake of a given element increases in proportion to the amount of that element added to the soil. Theoretically, a concentration eventually would be reached at which the elements would become toxic and hence inhibitory to plants. Chances are that this level would be reached before the concentration in the plants became toxic to animals consuming them. (2) The concentration of metals is greater in the leafy portion of the plant than in the seeds or roots. (3) Uptake of the metal is strongly dependent upon the pH level of the soil. Uptake is greatest in acid soils, less so in neutral soils, and least in alkaline soils. (4) Plants differ in their ability to accumulate trace elements in their tissues.

Benefits to Soil

The beneficial aspects resulting from spreading sludge on the land are similar to those to be gained from the use of compost, with the exception that most sludges have a greater fertilizer potential than the usual compost product. Fertilizer elements of importance to be found in digested sludges are nitrogen (N), phosphorus (P), and potassium (K). Total nitrogen concentrations range from 0.7 to 5.1 percent (dry weight); total phosphorus, 1.1 to 6.1 percent; and potassium, 0.2 to 1.1 percent.[95,96,114] Generally, activated sludge has a heavier concentration of these elements than does digested sludge. A comparison between the plant nutrient values of the two sludges and of animal manures is made in Table 8. As the table indicates, activated sludge and chicken manure are comparable to each other. Digested sludges surpass average farm manure in fertilizer value.

TABLE 8.   COMPARISON BETWEEN ACTIVATED AND
DIGESTED SLUDGES AND MANURES[97]

Item	N (%)	P_2O_5 (%)	K_2O (%)
Chicken manure	4.1	3.7	2.3
Av. farm manures	1.2	0.6	1.2
Av. activated sludge	5.6	5.6	0.4
Av. digested sludge	2.0	1.1	0.2

An appreciation of the fertilizer value of digested sludge can be gained from the following example: The particular sludge contains 3 percent N and is applied at 25 tons (dry weight)/acre (55.0 metric tons/ha). This results in a total loading of 1,500 lbs. N/acre (1,680 kg/ha), of which 150 lbs. is NH_3 and NO_3, and 1,350 lbs. (1,513 kg/ha) is organic − N. The latter will become available as it is converted to NH_3−N and NO_3−N, a process which may be spread over years. Thus the organic − N acts as a reservoir of nitrogen. This example also illustrates the need to take into consideration all of the not immediately available nitrogen in sludge when calculating a given nitrogen loading.

As with compost, the principal value of sludge in agriculture is in its improvement of the physical properties of the soil. In fact, the reactions ascribed to compost in soil also apply to those of sludge. To emphasize these benefits, they are summarized again as related to sludge: Sludge changes the soil structure by increasing the tendency of soil particles to aggregate and thereby enhance the water-holding capacity of the soil. As a result, plants are better able to cope with drought conditions. Heavy soils are lightened, i.e., made friable and loose. This enhances water penetration and thus minimizes ponding and runoff during rains. Gas exchange is aided, i.e., O_2 can move into the soil to the roots and CO_2 can diffuse outward. Thus root growth is encouraged.

Constraints

Despite the many benefits accruing from the use of sludge, certain constraints must be followed in its utilization in agriculture. These constraints find their expression mainly in terms of permissible loading both with respect to protection of public health and to maximizing crop yields. A major constraint is the presence of metals and certain toxic substances in sludge. A second constraint relates to the danger of excess nitrogen leaching to the ground water. A third major constraint pertains to the presence of pathogens. Since the first and second constraints were discussed in the section on loading, only that of pathogens receives attention in this section.

■ **Pathogens:** When it is encountered, the problem of pathogens is much more readily resolved than that of heavy metals. Before discussing remedies, it is well to speak of the nature and extent of the pathogenic aspects. Since allusion to the pathogen content of the various sludges has already been made in the section on the properties of sludges, this section concentrates on survival times in soil and on means of reducing the pathogen hazard.

The problem of pathogens has two aspects, namely, (1) types and numbers of organisms and (2) their length of survival in the soil. The significance of the first aspect is obvious. It determines the degree of hazard. The second is less direct, but it does bring in the time factor.[115] Its significance is that while a given sludge may not be safely used in one situation, in another it may. For example, it would not be prudent to use freshly processed digested sludge to grow a root crop; whereas there would be no danger in doing so in corn production. On the other hand, thoroughly aged sludge could be safely used for both applications. A difficulty in evaluating this second problem is one in common with heavy metals, namely, the diversity of opinions and of findings reported in the literature.

The observed survival times of a number of organisms in soil are listed in Table 9.[116] One of the better reviews on the influence of various factors on survival in soil to be found in the literature is that by Gerba et al.[117] According to their review, the factors of importance in survival in the soil, other than the genetically inherent resistance of the strain of the microorganisms, are moisture content, moisture holding capacity of the soil, temperature, pH, sunlight, concentration of organic matter, and antagonisms of and competition with other microflora.

TABLE 9. SURVIVAL TIMES OF VARIOUS PATHOGENS IN THE SOIL AND ON PLANTS[116]

Organism	Medium	Survival Time
Ascaris ova	soil	Up to 7 years
	vegetables	27–35 days
Salmonella typhosa	soil	29–70 days
	vegetables	31 days
Cholera vibrio	spinach, lettuce	22–23 days
	nonacid vegetables	2 days
Endamoeba	vegetables	3 days
histolytica	soil	8 days
Coliforms	grass	14 days
	tomatoes	35 days
Hookworm larvae	soil	6 weeks
Leptospira	soil	15–43 days
Polio virus	polluted water	20 days
Salmonella	radishes	53 days
typhosa	soil	74 days
Shigella	tomatoes	2–7 days
Tubercle bacilli	soil	6 months
Typhoid bacilli	soil	7–40 days

The effect of moisture on survival is exemplified by that on *Salmonella typhosa*. *S. typhosa* dies within four to seven days in dry weather in sandy soils and persists longer than 42 days in wet soils, such as loam and adobe peat. (Sandy soils have a very limited moisture-holding capacity.) *S. typhosa* and *Streptococcus faecalis* serve as exemplars for the effect of pH. They survive only a few days at a pH of 3 to 4 and several weeks at pH 5.8 to 7.8. The influence of pH is due both to its direct effect on viability of the organisms and to that on the availability of nutrient or interference with the action of inhibiting agents. Sunlight exerts its influence through the ultraviolet component of its radiation spectrum.

Pathogens survive longer at low temperatures. Some, such as *S. typhosa* may remain viable as long as 24 months at freezing temperatures. Indeed, the tolerance of *S. typhosa* for freezing temperatures is outstanding in that the organisms remain viable down to −45°C. Dysentery bacilli survive as long as 135 days at subzero temperatures. Fecal coliforms suffer 90 percent reduction in three or four days in the summer and in 13 or 14 days in the winter. Certain pathogens, and especially the parasites, are not as tolerant of low temperatures. In fact, a number are geographically limited because of their inability to remain viable at cold temperatures. Nevertheless, the persistence of the salmonellae and other dysentery-causing organisms at low temperatures is an important factor when considering the effect of time on the degree of a hazard in the use of sludges. The upshot is that storage would have to be longer in those regions having freezing winters.

The presence of organic matter in the soil prolongs the survival time of many pathogens, especially of the enteric microorganisms such as *Salmonella typhosa, Escherichia coli, Streptococcus faecalis,* and *Shigella flexeri*. This is due to their ability to metabolize certain of the nutrients in organic matter. This ability explains the occasional increase in number of certain pathogens following introduction into a soil.

The antagonistic action of other microflora can have a depressive effect on the survival of pathogens. This is especially apparent in soils possessing a sizeable actinomycetes population. Of course, competition is yet another important factor. The indigenous bacteria, being better adapted to soil conditions, are more successful in the competition with the introduced pathogens for available nutrients. Naturally, the antagonistic effect and competition are also influenced by moisture content, pH, and temperature of the soil.

The lesson to be drawn from the discussion of the factors affecting the survival of enteric bacteria and parasites in the soil is that no hard-and-fast time span can be named, at the end of which a given sludge is rendered safe for all uses. However, a year of storage would seem to preclude the continued viability of all pathogens likely to be found in sludges. The only exceptions would be those in which an unusual storage condition is coupled with exceptionally durable microorganisms. Two examples come to mind, namely, *Mycobacterium tuberculosis* in a shaded, dry soil and spores of *Bacillus anthracis* in dry soil shielded from sunlight.

At the time of this writing, little information was to be found in the literature on the survival of viruses in soil. Most of the studies to date have been with poliovirus 1. Depending upon pH, temperature, moisture, and the nature of the soil, survival time of this virus may range from 28 days to 170 days.

A means of ensuring the biological safety of a sludge is to pasteurize it at 70°C for 30 minutes. The disadvantage of this course is primarily one of economics, i.e., cost of the energy involved. Adding enough lime to raise the pH to 11.5 or higher is another possibility. The material is held at this level for two hours or more.

Applications

In general, it may be said that the applications suitable for treated sludge are the same as those named for compost. Thus the uses range from agricultural to land reclamation. Sludge may be applied either in the "liquid" (slurry) or in the dried form. The use of one or the other form is determined by the exigencies of a situation, and each has its advantages and disadvantages. Unless special provisions are made to retain the material, a sludge slurry should not be used on a hillside having a slope in excess of a 5 percent grade. Contour plowing with injection of the slurry in the furrows would extend the steepness of the permissible slope. Moreover, care has to be taken to avoid flooding or ponding. This would result in the development of nuisances and the promotion of undesirable anaerobic conditions in the soil.

Advantages in the use of sludge in its slurry form are the avoidance of special dewatering steps, the ease of application, and perhaps the facility of transportation. The advantage of not requiring dewatering and the consequent energy saving needs no further elaboration. Ease of application comes from the ability to pipe or spray the material on the fields. However, this advantage is lessened by the need for tank trucks to enter the fields. Spraying results in the formation of aerosols. The benefit of ease of transportation would be diminished if tanker trucks were used instead of pipes (hydraulic) to transfer the slurry from treatment plant to the site of application. The use of truck transport imposes the carting of water as well as of sludge. However, this latter apparent disadvantage is modified by the fact that the liquid phase contains a significant amount of nutrients that would be lost during dewatering other than by evaporation.

The use of sludge in the solid form has the advantage of permitting prolonged storage with a minimum occupation of land area and a relatively minor danger of nutrient loss. The only limitations imposed by topography are those of accessibility to the fields. The amount per individual application is not determined by danger of flooding or ponding. Transportation involves the carrying of solids only and hence results in a greater percentage of the useful component of sludge per truckload. For example, a 20-ton load of slurry would have only from 1 to 2 tons of solids, whereas the same weight of dried sludge (20 to 25 percent moisture content) would have 15 to 16 tons (13.5 to 14.4 metric tons) of solids.

Probably the most successful large-scale reclamation project at this time is being carried out in central Illinois. Here, some 1,500 acres (607 ha) of strip-mined land have been placed in production, with digested sludge serving as the sole fertilizer. In 1973 about 55,000 bushels of corn and 6,000 bushels of beans were produced. Digested sludge not only serves as a fertilizer, it also is a source of humus and acts as a cover for the raw land.[8]

The application for specific situations is limited by the constraints discussed in the preceding sections. An additional constraint is public acceptance or at least permission of those individuals whose residences are in the vicinity of the site destined to receive the sludge. When sludge was first applied in a large-scale demonstration operation in central Illinois, the opposition of the local citizenry was decidedly vociferous despite a preceding public relations campaign. A part of the difficulty was a combination of fear of and indignation at becoming a "dumping" ground for Chicago's sewage wastes. The popular concept of sewage did not help matters. This experience indicates that public fears must be allayed and that mere promises are not enough. A preliminary small-scale demonstration is needed. As more successful full-scale operations come into being, they will serve as proof that sludge can be disposed of on the land in an acceptable and innocuous manner. Of course, the neighbors must be assured that the entire operation will be carefully monitored. It goes without saying that this monitoring must be faithfully carried out. The monitoring should be based on those parameters that have a bearing on ground and surface water contamination and on soil fertility.

Economics

Because of inflation, any costs cited herein are of necessity relative. Moreover, because of rapidly changing conditions that affect economics, present relationships may not be the same in the future. Since energy costs and even energy availability are important factors, costs cited at this time are indicative rather than firm.

From 1963–1974, the heat-drying (1400°F or 760°C) cost was about $106/dry ton.[99] The greater part of this cost was for energy. In this particular case, the sludge was sold for $14 to $16/ton ($15.40 to $17.60/metric ton). Producing and transporting liquid sludge (4 percent to 6 percent solids) for recycling cost from $30 to $60/dry ton ($33 to $66/metric ton).[99,118] The least expensive form is air-drying — providing the necessary land area is available for drying beds. The costs in this case run from about $8/dry ton ($8.80/metric ton) and above to distribute it.[99] At the time of this writing, the most recent cost figures on hand for applying and using sludge were an estimated $150 to $200/dry ton ($167 to $222/metric ton) to recycle sludge on strip-mined land in central Illinois.[99] These figures include costs of land, barging charges, and the expensive operation of regrading the fields.

In another operation in which liquid digested sludge is being used as a fertilizer, a private company is selling lagooned liquid digested sludge to local farmers for $15/acre ($37/ha). However, it costs the metropolitan Chicago sanitation district $50/dry ton ($55/metric ton) to transport and lagoon the sludge. (The private company is under contract to empty the lagoons.) On the other hand, it should be remembered that the district must dispose of the sludge in one manner or another. The important point is that to dispose of it in a useful manner costs less than to incinerate it.

Composting Sewage Sludge

Composting can solve three major problems inherent in the disposal of sewage sludge on land by (1) stabilizing excess quantities of nitrogen that otherwise might percolate to the ground water, (2) killing disease-causing organisms, and (3) eliminating objectionable aesthetic characteristics associated with conventional anaerobic digestion of sludge. Moreover, when composted with another material, the concentration of metals in the sludge is thereby diluted. Another advantage related to dilution is that, in composting, carbon is supplied to permit the microbial assimilation of more nitrogen in the sludge than would be possible without composting.[119] An important limitation in nitrogen loading with sludge alone is an insufficiency of carbon for microbial metabolism. Yet another advantage is that composting broadens the variety of uses to which sludge can be applied by converting it into an easily workable and storable product suitable for large-scale farming and for the home garden.

The concept of composting sewage sludge is not a new one. In the early 1950s pilot-scale experiments involving the composting of raw and digested sludges with municipal refuse were conducted by the University of California (Berkeley).[16] During the mid-1950s, a successful commercial operation in which digested sludge was mixed with wood chips and composted was carried on in the San Francisco Bay area. Other studies in which sewage sludge was mixed with segregated municipal refuse and compost were conducted in the Los Angeles area. A fertilizer company in the latter area has been selling fortified sludge compost during the past several years.[120]

The most comprehensive study in the United States to date on the agricultural aspects of sludge utilization is one going on at the

U. S. Department of Agriculture Research Service experimental station at Beltsville, Maryland.[43]

■ **Methods:** A primary requirement for composting undewatered sewage sludge is that it be mixed with some material capable of serving as an absorbent and providing porosity to the mass. This is true because of the high moisture content of sludge (90 to 95 percent). Of course, if the sludge is dewatered (i.e. to less than 55–60 percent moisture) prior to composting, an absorbent material may not be necessary, although it would be advisable. In fact, dried or composted sludge can be mixed with the liquid sludge providing that the net moisture content of the mixture does not exceed 55 to 60 percent.[121] Fibrous materials or those having an appreciable "structural" strength make excellent absorbents. Examples include those named in a previous section for cannery wastes, namely, wood chips, sawdust, straw, dried grass, dry leaves, refuse, etc.

When sludge and refuse are composted together, obviously the refuse should receive the pretreatment described for composting refuse alone. The proportion of sludge to refuse is a function of the moisture content of the sludge. It has been estimated that the output of compostable refuse from an average city would accommodate about one-third of its output of digested sludge. Therefore, if the sludge were relieved of about two-thirds of its water, all of the sludge could be composted without difficulty, providing it could be thoroughly mixed with the absorbent. An important contribution by the sludge fraction is its nitrogen content. It is to obtain this nitrogen that prompts those who design plants for composting municipal refuse to include a provision for adding sludge. If wastes other than refuse are used as an absorbent, they are mixed with sludge in proportions that result in an appropriate moisture content.

In an operation involving an enclosed digester, the sludge can be sprayed on the milled refuse as the latter enters the digester. In windrow composting, the mixture can be sprayed on the refuse as it leaves the grinder or when it is stacked at the site of the windrows. In the latter case, the procedure followed in the U. S. Department of Agriculture (USDA) studies is a good one.[43] The absorbent is placed in a strip about 12 in. (30 cm) deep and about

15 ft. (4.5 meters) wide. Sludge (75 to 85 percent moisture) is distributed over the absorbent to an average depth of 4 in. (10 cm). The sludge and absorbent are then blended and formed into a windrow.

The principles applicable to composting in general are followed in the remainder of the process. In one of the approaches studied at the USDA's Beltsville station, a newly set-up windrow is turned over daily for 10 days. Then it is spread to a uniform depth of in a row 20 ft. wide and cultivated to air-dry over a period of two or three days. When the moisture content has dropped to 35 percent, fresh sludge is added and the mixture is then windrowed and turned for an additional two weeks. A Terex-74 automatic turner is used to turn the piles. At the end of the two-week period, the material is dried again and is screened to recover the wood chips for use. The screened sludge is stockpiled for a period of 30 days before it is considered ready for use in field or garden. Suitable variations of the procedure can be made to fit particular situations.

Recently, a new system of composting sludge that offers much promise was developed in the USDA's Agricultural Environmental Institute at Beltsville[122] and is being tried on full scale at Bangor, Maine. The system involves the building of a pile of raw sludge mixed with wood chips or shredded bark to provide bulk and the application of a vacuum to draw air through the pile. The pile is covered with a 30-cm layer of finely screened (passes a 1.0-cm screen) compost from a previous run. (If no compost is at hand, wood chips or shredded bark can be used.) The cover is to prevent odoriferous gases from escaping into the atmosphere. The gases removed from the pile are scrubbed by passing them through another (smaller) pile of screened compost. The arrangement is diagrammed in Figure 14.

In setting up a pile, a flexible perforated 10-cm drainage pipe is laid in a U shape on the ground. The pipe is covered with a 30-cm layer of unscreened compost or wood chips to absorb liquids seeping from the pile and to keep the pipes from becoming clogged. Then a mixture (1:3 volume basis) of filter cake raw sludge (23 percent solids) is placed to a height of about 2.5 to 3 meters over the pipe and its covering. The entire pile is then covered with a 30-cm layer of screened compost or with wood

AERATION PILE

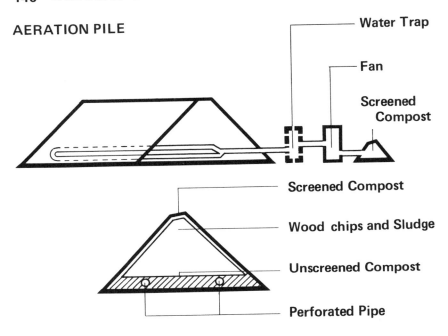

Figure 14: USDA Pile arrangement for composting raw sewage sludge

chips. The perforated pipe is attached by a piece of solid pipe and connectors to a blower.

Suction is applied over a 10-day period, which may be extended to 14 days. After that the blower is reversed and air is pushed into the pile for seven to 10 days. Reversal of the air flow results in an increase of temperature in the cold regions of the pile. No odors are formed because by the end of the 10 to 14-day period the greater part of the sludge has been fairly well stabilized. After 21 days, the compost is moved to a stockpile and cured for about four weeks.

Monitoring consists in following the course of the temperature rise and fall and in determining the oxygen content of the exhaust gases. The oxygen content should range from 4 to 15 percent.

To avoid the need for transportation or because of difficulty in obtaining waste material to serve as absorbents, it may be advisable to recycle them through several compost cycles. This is easily accomplished with the coarser absorbents by screening the dried compost mixture. The sludge itself composts to fine granular particles while the absorbent retains for the greater part its original physical dimensions. Screened or not, the final product is an excellent one not only in terms of appearance, but also as a soil amendment. Screened compost produced at the Beltsville plant averages about 0.9 percent nitrogen, 2.3 percent P_2O_5, 0.2 percent K_2O, 50 percent organic matter, 35 percent moisture, 27 ppm boron, 0.04 percent sulfur, 0.3 percent magnesium, and 2.6 percent calcium.[43,122]

Public Health Aspects

Those constraints pertaining to composting in general also apply to composted sludge. Proper turning is especially important because at one time or another all material must be exposed to the high temperatures prevailing in the interior of the pile. If a mechanical turning device is used, the position of the exterior layers and interior layers is not necessarily interchanged at each pass of the machine. Instead, a randomized mixing occurs. Under such circumstances the required frequency of turning becomes greater to ensure the proper exposure of all the material. The problem is complicated by the fact that in the mixing a sterilized portion can be re-inoculated by organisms from unsterilized portions. However, the Beltsville studies indicate that the survival of pathogens is not great enough to constitute a health hazard. The reasons are that as composting proceeds, the material apparently becomes less capable of supporting the growth of pathogens, and the heat taking place during the 30 days in the stockpiled material reduces any remaining pathogens to insignificant numbers. Studies conducted thus far indicate that, providing high temperatures are reached throughout the pile, pathogen die-off in the modified forced-air system parallels that in turned windrow systems.[117,123]

Despite the potential die-off of pathogens, the composting of raw sewage sludge is not recommended.[124] In addition to the public health hazard, there is the very likely occurrence of highly objectionable odors. In studies conducted by this author, despite a very successful experience with composting mixtures of raw sludge and refuse, an odor not unlike that of an abandoned privy was always noticeable in the vicinity of the piles, even in the presence of thoroughly aerobic conditions. The presence of the usual earthy odor and of prevailing winds off San Francisco Bay could not hide the objectionable odor. Other dangers listed in the section on composting in general also apply here. However, the Beltsville experience indicates that with their forced-air system the preceding difficulties can be minimized.

Economics

The cost of composting sludge certainly is much less than that of incineration. In the absence of other reported numbers, those cited on the basis of the USDA Beltsville study[43,122] are given here. With the turned windrow system, the processing cost at Beltsville is on the order of $3.50/ton ($3.88/metric ton) wet sludge (20 percent solids) for a 600-ton/day (540 metric tons/day) output. This cost does not include that of transporting sludge to the site and of the land.

Approximately 30 acres (12.15 ha) of land would be required for a 600-ton/day plant. This would meet the area requirements for the compost pad, storage and administrative areas, roadways, and surface runoff control. Stabilization of the storage area and compost pad can be in the form of crushed stone on a compacted earth sub-base. While the actual dollar costs vary with time, place, and nature of the operation, the numbers arrived at in the Beltsville study are useful for making comparisons and arriving at an approximation of the costs to be expected. Based on a 600-ton/day plant and following the procedure at Beltsville they are:

I. Investment
    Total investment $1,718,880
        (equipment and construction)
    Annual cost
    Total capital cost   326,970

II. Operating Costs

Wood chips	54,000
Labor	266,000
Fuel and oil	41,610
Compost analysis	40,000
Electricity	2,400
Total operating	$404,010
Total annual	$766,980
Cost per ton of wet sludge	$3.50

If the sludge were disposed by incineration, the annual cost for the oil alone required to provide conditions for combustion of the sludge cake (82 percent solids) would be on the order of $958,000 as based on a cost of $0.27 to $0.32/gal. oil or $17.50/metric ton dry solids.[123]

The cost of composting sewage sludge with the USDA modified forced-air systems is much less expensive than that with their turned-windrow system. Although all of the costs have not been determined as yet, an idea of the total required expenditure may be gained from the costs of the pieces of equipment and material required to process 50 tons of 23 percent filter cake (10 dry metric tons) as listed by Epstein and Willson.[122]

Expressed as 1975 dollars, they are: (1) 90 ft. (27 m) of 4-in. (10 cm) perforated plastic drainage pipe = $24.00, (2) 40 ft. (12 m) of 4-in. solid plastic pipe = $11.00, (3) one 0.33-h.p. 5.3 to 6.6 amp. motor with blower = $73.37, (4) two 0.25-in. (1 cm) plastic straight connectors = $0.70, (5) one 4-in. (10 cm) plastic "T" connector = $2.65, (6) timer = $16.87. Additional costs would be for temperature and oxygen probes, a front-end loader, and possibly a rototiller. The latter two are used for setting up the windrows and mixing the wood chips and sludge.

Land requirements for composting the 50 (wet) tons of sludge cake are about 800 sq. ft. (74 m²) for a windrow approximately 20 ft. (6 m) wide and 8 ft. (2.5 m) high. Additional area is needed for maneuvering equipment and constructing other piles of the daily increment of sludge that is produced during the three-week composting period. An area is needed for curing the material and for screening, drying, and storage. Land area would have to be acquired for administration, parking, roadways, etc. Epstein and Willson estimate the total land area for an 11-dry-ton/day (10 metric ton/day) operation to be about 2.47 acres (1 ha).

The preceding cost figures show that, from a purely economic point of view, composting digested sludge has a great advantage over incinerating it, especially if experience proves the modified forced-air system described by Epstein and Willson to be satisfactory in terms of aesthetics and public health. The economic advantage with the modified forced-air system would be magnified enormously if its operation and its product comply with all of the rigid safeguards needed when dealing with raw sewage sludge.

It should be noted that if land costs or other factor dictate the siting of the compost operation at an appreciable distance from the sewage treatment plant, the cost of transporting the sludge to the site would become a major item.

3. ANAEROBIC DIGESTION

INTRODUCTION

Anaerobic digestion is a term commonly applied in waste treatment to a process in which the wastes are "stabilized" through biological activity in the absence of atmospheric oxygen with the concomitant production of methane (CH_4) and carbon dioxide (CO_2). In the broad biological sense, the production of methane is not an essential element in the definition of *anaerobic digestion*, and the term *fermentation* may be used instead of *digestion*. The definitive phrase is "absence of atmospheric oxygen." The history of the utilization of anaerobic digestion in the treatment of wastes practically parallels that of wastewater treatment in that it is used to stabilize the sludges produced in the treatment process. However, in its first stages of technology, anaerobic digestion was rather primitive and took the form more of a "holding" tank than of a unit designed to control and facilitate the fermentation process, as are modern digesters.

The anaerobic digestion of farm wastes also has a relatively long history, but it never was practiced on a scale of any magnitude. It was used on a few farms in France and Germany during World Wars I and II, not so much as a waste treatment device, but rather to produce the combustible gas CH_4 for use a a fuel for household cooking.[34] It has been practiced for the past 30 to 40 years in rural India, at first on a pilot level and lately developing into a meaningful scale — more than 1,500 installations by the later 1960s.[125] In the Indian application, both human and animal excreta are digested.

The recently realized shortage of natural gas and of fossil fuels in general in the United States has spurred the development of several animal-waste treatment operations in the design of which production of CH_4 ranks equally in importance with stabilization of the manure.[126] Except for a brief interlude in the 1930s, anaerobic digestion of municipal refuse received scant attention both on a research level and on a practical scale until the later 1960s. The spurt in the 1930s was a response to the introduction of the home garbage grinder. Professionals responsible for waste-water treatment, specifically of primary sludge, were concerned over the possibility of adverse reactions of the digester culture to the introduction of kitchen wastes, and over the amount of the increase in organic loading to the digester. Consequently, a period of research activity on the "digestibility" of the garbage fraction of refuse took place.[127]

Other than the kitchen garbage flushed into the sewers by the home grinders, large-scale digestion of the garbage fraction has been limited to only a couple of installations.[128-130] Even in these installations the garbage was digested along with sewage sludge. It was not until the late 1960s that the possibility of digesting refuse as a solid-waste treatment process received serious attention, and then only as a means of disposing of hydraulically transported refuse. As such, it would be a slurry form and thus at least physically amenable to conventional anaerobic digestion. The activity at the time consisted of a study aimed at determining the effect of the components of refuse individually and *en masse* on the performance of a conventional sewage sludge digester.[131,132]

As was the case with animal wastes, the recent shortage of fossil energy sources has generated a considerable amount of interest in the production of methane through the anaerobic digestion of organic refuse. However, while the interest has led to extensive investigations,[126] it will be some time before anaerobic digestion is practiced on a municipal scale. In this presentation, the anaerobic digestion of sewage sludges is not included because the subject is thoroughly covered in existing sanitary engineering texts.[90,133] Instead, major attention is directed toward agricultural wastes and the organic fraction of municipal refuse.

PRINCIPLES

The Process

The anaerobic digestion process is characterized by two phases, popularly termed the "acid phase" and the "methane phase" (or synonyms thereof). The basis for the terminology is the sequence of events that typically take place in a digester. At the start it should be emphasized that this sequence is recognizable as two distinguishable steps only in a batch operation or at the initiation of a continuous one.

The acid phase is the first and is distinguished by the breakdown of complex organic carbonaceous materials (proteins, carbohydrates, fats, etc.) by acid-forming (i.e., nonmethanogenic) bacteria into organic acids (principally short-chain fatty acids), CO_2, H_2, NH_4^+, and H_2S. In the methane phase, methane-forming bacteria convert fatty acids, CO_2, and H_2 into CO_2 and CH_4. The production of CH_4 by the methane formers takes place through the fermentation of short-chain fatty acids and some alcohols and through respiration involving the anaerobic oxidation of H_2 and CO and simple organic compounds. The oxidation is coupled with the reduction of CO_2 to CH_4. The overall process is nicely illustrated by the schematic description of cellulose developed by Chan[134] and given in Figure 15. Since cellulose is a major constituent of most solid wastes, the description is especially apropos. An example of the fermentation reaction is the decomposition of acetic acid and methyl alcohol:

(acetic acid)
$$CH_3-COOH \longrightarrow CH_4 + CO_2$$

(methyl alcohol)
$$4CH_3OH \longrightarrow 3CH_4 + CO_2 + 2H_2O$$

The incomplete oxidation of alcohol to acetic acid, coupled with the reduction of CO_2 to CH_4 as carried out by *Methanobacterium omelianski,* is an example of production of CH_4 through respiration. It is

$$2CH_3-CH_3OH + CO_2 \longrightarrow 2\ CH_3-COOH + CH_4.$$

The reduction of CO_2 with molecular hydrogen is thus:

$$4H_2 + CO_2 \longrightarrow CH_4 + 2H_2O.$$

Rate-Limiting Factors

From the foregoing description of the process, it becomes apparent that four steps have the potential to become rate

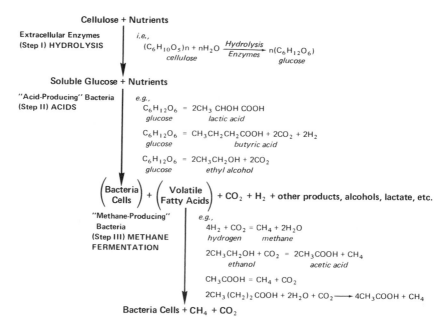

Figure 15: Schematic description of cellulose anaerobic fermentation[134]

limiting. They are (1) the conversion of insoluble cellulose by extra-cellulases into soluble carbohydrates (e.g., glucose), (2) the activity of nonmethanogenic bacteria in converting carbohydrates into low molecular weight fatty acids (predominantly acetic and proprionic), and (3) conversion of the acids by methanogenic bacteria into dissolved CH_4 and CO_2. Recently, a fourth limiting step has been proposed: (4) transfer of the dissolved products from the liquid to the gas phase.[135]

The acid phase proceeds much more rapidly than the methane phase and is much less susceptible to upset from adverse conditions. The bacterial population is highly diverse in composition and probably is common to all anaerobic fermentation in terms of types, if not in concentration of individual types. The acid formers are responsible for the initial drop in pH that is characteristic of anaerobic digestion.

The methane bacteria are characteristically slow growing and are quite fastidious in their requirements. For this reason, the methane stage has generally been regarded as the limiting step in the digestion process as a whole. Moreover, it is a limiting factor not only with respect to rate of destruction (i.e., stabilization) but also to the continued survival of the entire microbial population. The reason is that continued utilization of the volatile acids formed in the first step prevents an accumulation of the acids to a bactericidal level, i.e., "ensilaging."

Finney and Evans,[135] the proponents of step 4 as a potential rate-limiting factor, reason that as the biological retention time (SRT) is shortened, i.e., development period lowered, individually the bacteria must increase their rate of reproduction. This latter in turn leads to an increase in gas production and transport requirements. If transport of gaseous products away from the individual bacterium were indeed rate limiting, then a reduction in SRT would lead to an increase in acid concentration. The inhibition resulting from the phase transfer of products (i.e., from liquid to the gas phase) may arise when gas bubbles completely surround the individual bacterium, as would occur at very high substrate concentrations. The "wall of bubbles" would interfere with the diffusion of substrate into intracellular spaces. The authors maintain that they proved their point in experiments involving very vigorous agitation of the culture in a low-pressure environment. Regardless of the validity of their reasoning, the intensive mixing would promote the efficiency of the process.

Environmental Factors

■ **General:** In addition to the four potentially rate-limiting steps, the efficiency of anaerobic digestion depends upon an array of environmental factors that affect all biological systems. With respect to design and operation of a digester, the more important of the factors are temperature, hydrogen ion concentrations, and substrate. Anaerobic digestion differs drastically from composting in terms of oxygen requirement. Whereas modern composting involves a need for oxygen, in anaerobic digestion oxygen is toxic and hence inhibitory. The maintenance of a low oxidation-reduction potential is a requisite for the continued production of methane. Therefore, not only oxygen, but also any highly oxidized material (e.g., NO_2, NO_3) can lead to the inhibition of methanogenic bacteria.

● *Temperature* — As in all biological processes, temperature exerts an important influence on the rate and efficiency of the digestion process. Measured in terms of gas production and volatile matter destruction, a straight-line increase in efficiency parallels a rise in temperature from ambient to 30 to 35°C. Thereafter, the increase begins to level off and reaches its maximum at a point between 35 and 40°C. A slight drop occurs within the 40 to 45°C range; but as the temperature rises above 45°C, efficiency again increases and peaks at about 55°C. Of course, the culture has to be adapted to thermophilic conditions before the increase above 45°C can take place. Without the adaptation, the culture would be killed off. Activity declines drastically at temperatures above 65°C. The effect of temperature on gas production is indicated by the relation of the curves in Figure 16.[136]

There is some argument over the virtues of thermophilic vs mesophilic digestion. For practical purposes, mesophilic digestion perhaps has the advantage, despite a slight superiority in efficiency and in characteristics of the sludge at thermophilic temperatures. The latter two advantages of thermophilic digestion are offset by the significantly greater energy required for maintaining the

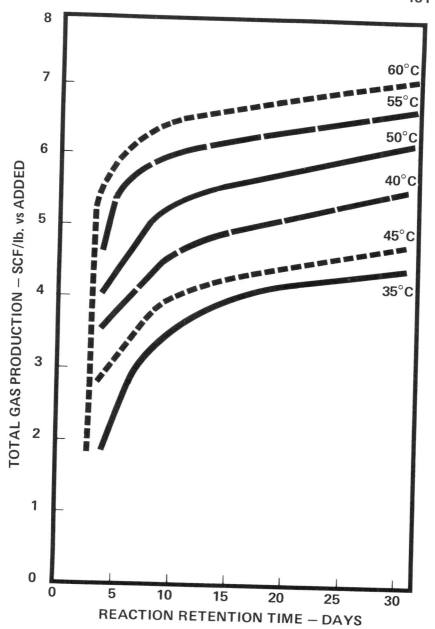

Figure 16: Effect of temperature on gas production

digester culture at the high temperature. Moreover, if the culture should be killed off through some mischance, it would be necessary to adapt a new culture, a process that could require a month or more.

• *Hydrogen Ion Concentration* — The pH level of the digester culture is an important factor because of the almost acute sensitivity of methane bacteria to an unfavorable pH level. The optimum range for the methane producers is between 7.0 and 7.2, although digestion proceeds quite well between pH 6.0 and 7.6. The inhibitory effects become apparent at levels either higher or lower than the 6.0 to 7.6 range. Activity ceases when the pH drops below 5.0 or rises above 8.0.

• *Substrate* — Broadly speaking, any waste that can be composted can also be anaerobically digested. However, as with composting, not all are as readily amenable to digestion. Generally, manures and sewage sludges are easily digested. Other highly organic wastes (e.g., cannery wastes and packing plant and rendering plant wastes) also lend themselves to digestion, providing the C/N ratio has been suitably adjusted. The preceding types of wastes are characterized by a certain degree of homogeneity, a high percentage of easily decomposable compounds, and a relatively small particle size. The solid waste mixture that constitutes municipal refuse, on the other hand, is not readily amenable to conventional anaerobic digestion and must be subjected to a considerable amount of processing to make it so. For example, it is an extremely heterogeneous mixture both physically and chemically. This heterogeneity applies to resistance and decomposition, as well as to suitability for processing. It complicates the design process with respect to both construction and operation.

The reasons for the relative ease or difficulty arising from the source, and consequently type of substrate, become apparent from the following discussion of the various factors pertaining to substrate:

In common with other biological processes, particle size is a major factor. In keeping with the rule that the greater the exposed surface area per unit of mass, the more rapid the breakdown, in anaerobic digestion the optimum particle size is the smallest one attainable with existing size-reduction technology. The importance of size is in direct proportion to the resistance of the material to biological attack. Thus, whereas a particle size of 1 to 3 cm would be satisfactory for green vegetable matter, the maximum for wood is less than a few millimeters.

Inasmuch as conventional digestion is based on a suspension of solids in water, solid wastes must be slurried. The solids content of the input slurry (feed) should be such that the solids concentration of the digester culture remains at about 5 to 8 percent. Thus, if 50 percent of the incoming solids are converted to gas, the solids concentration of the feed could be as much as 16 percent, and even greater if the rate of conversion to gas is increased proportionally. Of practical significance is the fact that not only is the yield increased with increase in efficiency, but also the required size of the digester for a given feed rate and operating conditions are correspondingly reduced. If the digester culture is too thick, it may become difficult to adequately mix it, and even to handle it. Too low a solids concentration involves processing a larger than necessary volume of material, and hence a lowered efficiency of digester utilization.

Carbon-nitrogen ratio (C/N) in the substrate exerts a decisive influence on the balance between acid formers and methane producers. If the C/N ratio is in excess of 25 or 30:1, nitrogen becomes a limiting factor. Since the acid formers are the more vigorous of the two groups, they multiply at the expense of the methane producers. The imbalance is manifested by an increase in volatile acid concentration. Generally, if the carbonaceous material or a significant fraction thereof is in a form resistant to bacterial attack, the permissible C/N ratio may exceed 30:1 without inhibiting the culture.

In the University of California studies[131,132] it was observed that when wood (sawdust), straw, and newsprint served as the major part of the substrate, the C/N ratio could be as high as 45 to 50:1 without exerting an unduly adverse effect on digester performance. If the C/N of a waste is too high, as would be the case if the entire organic fraction of (United States) municipal refuse were digested, the problem can be met by adding sufficient raw sewage sludge or an animal manure (e.g., poultry, swine, sheep) rich in nitrogen to the waste to lower the C/N to 25:1. The use of raw sludge as an additive has the advantage of providing for the disposal of that material.

Resistance of the material to microbial attack limits the rate of digestion and, at times, its efficiency. Obviously, if the substrate can be broken down only slowly or incompletely, less nutrient will be available to the bacterial populations. Thus, while all organic materials can be anaerobically digested, the required time to do so with many substances would make the process impractical for their treatment. For example, wood (sawdust and wood chips) is only slightly affected in a digester operated on a 30-day detention period. The primary reason is the difficulty of breaking down the lignin molecule. Inasmuch as a detention time longer than 30 days would make an already costly process more expensive, digestion of untreated wood is economically (monetarily and energetically) unfeasible.

The degree of digestibility can be increased by subjecting the material to physical and chemical conditions that result in the hydrolysis and partial breakdown of some of the material. [136-138] The treatment may involve the application of heat (150 to 200°C), pressure (125 to 220 psig), and a strong acid or strong alkali. For example, Gosset and McCarty[137] found in one experiment that with no pretreatment, gas production was 80.5 mg gas/gram of C.O.D. introduced; but with pretreatment, including exposure to a temperature of 150°C, it was 125 mg/gram C.O.D.; at 175°C, 135 mg/gram; and at 200°C, 151 mg/gram. In one study underway at the time of this writing, the plan was to restrict the treatment to the sludge produced from the digestion of untreated material. By so doing, the amount of material to be exposed to the acid or alkali treatment would be reduced. While pretreatment does materially increase the digestibility of the substrate, and thereby the amount of gas produced, the economic feasibility remains to be demonstrated; and, on first glance, it seems to be rather dubious. Moreover, there is the matter of the energy balance to be reconciled.

Availability of a substrate for microbial assimilation is more than a matter of susceptibility of the molecules of the substrate components to enzymatic reaction, it also is one of proximity of the bacterial cells. Proximity is ensured by mixing the culture.

Mixing not only makes new material accessible, it also serves to prevent or at least dilute the accumulation of bacterial waste products (i.e., metabolic wastes) in the immediate vicinity of the cells responsible for their formation. (Methods for mixing are discussed under "Design Considerations.")

Yet another factor related to substrate is loading. Loading determines the amount and rate of substrate introduction into the digester, and hence the availability of nutrients to the microorganisms. Insufficient loading results in a shortage of nutrients, and therefore a decline in digester performance. It also implies inefficient utilization of facilities. On the other hand, excess loading can lead to an imbalance in relative activities of acid and methane formers — and unfortunately in favor of the acid formers. The manifestation of the imbalance is a surge in volatile acid concentration and usually a drop in gas production. Another potential effect is an accumulation of toxic end products that in normal loadings would remain at a level not inhibitory.

The amount of permissible loading is a function of the density of the active microbial population, of the nature of the material, of the operational factors (i.e., approximation of optimum environmental conditions), and of the structural design. A suitable loading for animal manures ranges from 0.1 to 0.3 lb. volatile matter/cu. ft. culture volume/day (1.62 to 4.86 grams/liter/day) and for organic refuse (entire fraction) from 0.08 to 0.25 or 0.30 lb. volatile matter/cu. ft. culture volume/day (1.05 to 4.05 or 4.86 grams/liter/day).

● *Time* — A final factor to consider is time. While strictly speaking time probably should not be termed an environmental factor, it nevertheless is of major importance. Another complication in categorizing time is that it also enters into loading of continuous cultures. Time determines whether or not the final bit of energy or nutrient is taken from a substrate. It also determines whether or not the necessary microbial population can develop to the maximum density permitted by the environment and, in a continuous culture, whether or not it can maintain itself. If sufficient time is not allowed, the culture is continually diluted until it has been entirely "washed away." The optimum and the required amount of times are functions of the growth rate of the microbial populations. Since the methane formers are the slower growers in anaerobic digestion, their reproduction time becomes the limiting factor.

With conventional anaerobic digestion operated on a continuous basis, time is determined by adjusting the hydraulic loading rate by way of introducing a given volume of feed into the culture during the course of each day and withdrawing an equal volume of digester contents. This continuous adding and simultaneous withdrawal per unit of time is designated by the terms *detention* or *retention period* and is expressed in some unit of time. The culture can be operated such that, through the interposition of a settling chamber, separate detention periods can be applied for the solids (bacteria) and for the liquid phases. Except in very large operations, digesters are rarely operated on a strictly continuous basis, i.e., slurried wastes added continuously throughout the 24-hour period, day in and day out without interruption. The reason is to be found in the very small dosages indicated for small operations (e.g., individual farms). For example, a 900-gallon (3.4 m³) digester operated on a strictly continuous basis at a 30-day detention period would receive 30 gallons (113.6 liters)/24-hr. period, or a little more than 1 gallon/hour. Accurately metering even a clear liquid at that rate or even twice the rate (i.e., 15-day detention period) would be feasible only under laboratory conditions. Therefore, a compromise is made but always with the aim of approaching true continuity to the closest extent that is feasible. The greater deviations should be compensated for by corresponding prolongations of detention time. The reason is that the less frequently the new material is added and the treated material removed, the greater will be the depletion of the culture volume (and therefore of bacterial population) at each loading. Time must be allowed for the microbial population to regain its former density.

In a truly continuous culture the hydraulic loading rate (and hence detention period) is adjusted to precisely match the growth rate of the slowest growing microflora essential to the process, namely, the methane formers. Complex models have been developed for predicting rate of microbial growth, but they are all based on cultures grown either in defined or on relatively simple media and under precisely controlled conditions. Because of this background, the utility of the models cannot be compared with that of operational experience as far as the fermentation of a material as complex as municipal refuse or even manures is concerned.

SYSTEM PARAMETERS

Indices of digestion, i.e., system parameters, proven by years of experience to be useful in judging digester performance in the digestion of sewage sludge, are equally of value with respect to the digestion of other wastes. The major parameters are based on gas production, destruction of volatile matter, alkalinity, volatile acid content, and pH. Deviations from the observed normal pattern of these parameters in a given operation rarely are simultaneously abrupt and pronounced. When such a combination occurs, it is "catastrophic" and is in response to equipment failure, to gross mismanagement, or to the introduction of a very toxic material. More likely, deviations for the worse will be gradual and in the form of a trend. Such a trend serves as a warning of impending failure unless measures are taken to improve conditions. It should be emphasized, however, that it is an unusual operation that is not characterized by a certain amount of daily fluctuation in the values of the parameters. The result is that a plot of daily readings almost always takes the form of a serrate curve. The deviations beyond this band of daily fluctuations are the ones of significance.

Gas Production

Of the parameters, gas production is the most conveniently determined. Inasmuch as gas is a principal product of bacterial activity in anaerobic digestion, the volume of its production is an accurate measure of the extent of that activity. Since gas is produced through the breakdown of organic material, a large volume of gas production is indicative of an extensive breakdown of the material. Hence the efficiency of the culture as a whole in terms of stabilization of organic matter can be measured by the amount of gas produced per unit weight of organic ("volatile") matter introduced into the digester, usually expressed as cubic feet of gas (standard conditions) per pound (liters/kg) of volatile matter introduced. "Volatile" matter is practically synonymous with organic matter in wastes. (It refers to that fraction destroyed in the combustion of a sample.)

Inasmuch as some wastes are broken down more readily than others, the amount of gas varies accordingly. A few examples of reported gas production are listed in Table 10. When measuring the gas production efficiency in a new operation involving a substrate different from that to which the starting culture had been exposed, one should allow ample time for the organisms to adapt or for the enrichment for a suitable population to take place. Other than under laboratory or closely controlled cultural conditions, this may take from 30 to 90 days. Gas production also is expressed in terms of volatile matter destroyed. This measure has less significance as a parameter than volatile matter introduced. Its value is in predicting theoretical gas production from a given waste.

Since methane production is, as it were, the "capstone" of digestion as an energy production process, the total gas should contain a percentage of methane grossly paralleling that demon-

TABLE 10.   GAS PRODUCTION FROM THE DIGESTION
OF VARIOUS WASTES

Waste Source	Gas Production/Day[a] (cu. ft.)	(liters)	CH_4 (%)	Total Heat Value (Btu/Day)	Digester Volume per Animal (cu. ft.)	(liters)
Human	1.3[b]	36.8	66-75	900	1.0-4.0	28.3-113.2
Swine	6.3[b]	178.3	55-75	3600	1.0	28.3
Dairy cow	42.0[b]	1188.5	60-80	25000	37.0	1047
Poultry	0.4[b]	11.3	60-80	250	0.37	10.5
Municipal refuse	7.7[c]	480	55-60	4400	15-20[d]	1060-1247[e]
Green garbage	8.8[c]	550	55-60	4400	15-20[d]	1060-1247[e]
Garden debris	6.2[c]	410	55-60	4400	15-20[d]	1060-1247[e]

[a]Gas production per animal
[b]Reference 125
[c]References 131 and 132
[d]Volume per pound of dry volatile matter introduced per day
[e]Volume per kilogram of dry volatile matter introduced per day

strated as being characteristically generated by methane formers. Once the ratio of CO_2 to CH_4 has reached a plateau, that ratio becomes the measure or parameter for successful operation. As with efficiency, it may take some time before this level is reached with a new culture. Generally, the proportion of methane is greater than that of CO_2. With sewage sludge as the substrate, the CH_4 content can be as high as 75 percent, but more usually it constitutes from 60 to 70 percent. With other wastes, the percentage is on the order of 55 to 65 percent. Gases other than CO_2 and CH_4 are found only in minute amounts in digester gas. Among them are H_2 (<1 percent), H_2S, and NH_4, and of course, water vapor. Almost without exception a malfunction is predicted by a persistent rise in the CO_2 content of the gas and a corresponding drop in CH_4. This occurs because, being more resistant, the acid formers are inhibited less rapidly than the methane formers by adverse conditions.

Volatile Solids Destruction

If the objective of the operation is mainly waste disposal, then the destruction of organic matter becomes a key parameter. The destruction is limited to organic matter inasmuch as digestion is a biological process. Probably, net change or loss in volatile matter is a better way of expressing the parameter, because as used in practice it is a measure of the difference between the volatile matter of the solids introduced and that of the solids discharged from the digester (i.e., the residue). Destruction of the introduced matter is more complete than that indicated by the change or loss, because a sizeable amount is converted to microbial cellular material. The "loss" is in the form of carbon converted to CO_2 and CH_4. The loss in volatile matter varies with the resistance of the substrate to bacterial attack.

Volatile Acid Concentration and pH Level

Perhaps next to gas production and composition, volatile acid concentration ranks in importance as a practical parameter of digester performance. Both serve as warnings of incipient malfunction before a problem becomes serious enough to halt the process. The hydrogen ion (pH level) is not as sensitive as an indicator, as will be seen later, because the pH level may continue to be at a permissible level for some time after the volatile acid and gas parameters had begun to warn of imminent disaster.

The digester contents always have a detectable volatile acid concentration. In sewage sludge digestion it usually is on the order of 200 to 1,000 mg/liter. This level represents the equilibrium between the activity of the acid formers and the utilization of the acids by the methane formers in a given situation. It does not indicate the maximum level tolerated by methane formers. Actually, the concentration *per se* is not inhibitory: it is the combination of secondary effects resulting from the concentration. An example is the drop in pH level. The drop exerts an influence on the availability of the key nutrient elements and on the physiology of the bacteria. An everyday proof of the tolerance of the methane formers for volatile acids themselves is the relatively heavy concentrations of volatile acids used in media for culturing them.

This author has "adapted" continuous cultures to an input of 10,000 mg acid/liter of culture administered as a single dose each day. Acids tried individually were formic, acetic, propionic, valeric, and caproic. Gas production was in direct proportion to the quantity of acid fed the digesters. This experience serves to emphasize the point that concentration *per se* does not constitute the problem. The reason for the tolerance in the experiments was that a methane bacterial population was built up to a size commensurate with the acid dosage.

Volatile acid concentration as a symptom of malfunction takes the form of a sudden or persistent increase in concentration after an equilibrium has been reached or the failure to reach an equilibrium. The latter would occur at the initiation of a culture and would manifest itself by a straight-line increase in concentration with time. The increase after equilibrium indicates that some condition is inhibitory to methane formers and that, unless it is remedied, the culture will fail. In sewage treatment practice, an upper level of about 2,000 mg/liter is regarded as the danger point. However, in view of that which has been said, waiting until that point has been reached before taking remedial action would be poor management.

A difficulty in utilizing volatile acid concentration as a routine parameter is that of making the analysis. The analysis is a somewhat involved operation that readily lends itself to experimental error.

Normally, change in pH level could be taken as a measure of change in acid concentration. However, with respect to successful anaerobic digestion of the usual wastes, the cultures are so well buffered that any change in volatile acid concentration must be quite drastic before it is reflected by a change in pH level. The resulting slowness of its response lessens the utility of pH as a parameter. Contributing to the buffering capacity ("alkalinity") are a bicarbonate complex, the organic acids, and the acid salts. If used as a parameter, then a pH range of 6.5 to 7.5 is regarded as the optimum one. (As noted before, this is the range for the methane formers.) Departures from this pH range at either extreme indicate the development of conditions toxic especially to the methane formers.

Design Considerations

Design parameters for large-scale applications are determined by the type of digestion system to be utilized. At present, two systems, the standard-rate and the high-rate systems, are the most widely used in the United States. The principal distinction between the two is the absence of mixing in the conventional or standard-rate system.

As a result of the lack of mixing, the contents of the digester in conventional rate digestion become stratified into a gas section and a liquid section. The liquid section is stratified in the following order (from top to bottom): (1) a scum layer, (2) a relatively "clear" supernatant layer, (3) an active layer in which the major decomposition takes place, and (4) a stabilized layer at the bottom. Supernatant usually is returned to the head of the plant. In the digestion of solid wastes, the supernatant can be used as a slurrying medium.

Two problems generally characteristic of conventional-rate digestion become especially troublesome when processing municipal wastes and any waste that has a heavy concentration of wood or buoyant material. The first problem is the accumulation of fines. The second, and the more serious one, is the excessive buildup of the scum layer. Unless this layer is broken up, the greater part of the buoyant material finds its way into it and for practical purposes is no longer subject to decomposition. Moreover, a very thick layer interferes with the mechanical operation of the digester.

Because of the applied mixing, the high-rate process theoretically should be devoid of stratification and its attendant difficulties. On the other hand, because the stabilized and nonstabilized constituents are in a mixed state, a second process step is necessary. The second step may be either the adoption of a two-stage system or the provision of a settling tank which may be preceded by a vacuum degasifier.

In the two-stage system, the conventional- and high-rate processes are combined. This is accomplished by operating two digester units in series. Raw material is introduced into the first unit, the contents of which are mixed continuously. This unit usually is heated and most of the stabilization takes place in it. The discharge from the first unit is into the second unit, the contents of which are not mixed. Hence, stratification takes place in it, as in conventional-rate digestion.

A fourth type of process is beginning to come into usage. It is the anaerobic contact process. The contact process is similar to two-stage digestion and differs mainly in that sludge from the second stage is returned to the head of the first stage. The latter is analogous to the return sludge step in the activated sludge process.

While the contact process may work well in the digestion of raw or activated sewage sludges and of manures devoid of bedding material, difficulties could be encountered in applying it to the digestion of refuse. The sludge from digesting refuse contains a significantly large fraction of "inert," nonbiological materials. Inasmuch as the purpose of sludge recirculation is to assure the presence of a microbial population adequate to handle the loading to the system, recirculating refuse-derived sludge would therefore only partly accomplish the purpose of sludge recirculation. Moreover, handling problems would arise because of the nature of the nonbiological materials.

■ Operational:

● *Starting a Culture* — No difficulty is encountered in developing the populations necessary to bring about the acid phase. On the contrary, the problem is one of restraining the activity of the acid formers long enough to allow the buildup of methane bacteria. Consequently, starting a digester is mainly a process of developing a population of methane formers. The most rapid and reliable method is "mass inoculation" ("seeding") with actively digesting material. The volume of the "inoculum" should be about 10 percent of the volume of the total culture. A nearby satisfactorily performing digester is the best source of "seeding" material. Bottom mud from marshes is another, albeit less desirable, source. If no on-going digester is at hand, then one must rely on time and the chance presence of a few or more methane formers indigenous to the wastes. It has been the experience with sewage-sludge digestion that if conditions are appropriate, a sufficiently large population can be developed within 30 to 60 days without seeding.

● *Factors* — Inasmuch as operational parameters determine structural design, especially in relation to volume and accessory equipment, their discussion precedes that of structural design. The objective of operational design factors is to ensure an operational procedure that provides an optimum environment subject to the constraints of practical and economic feasibility. The "dimensions" named in the section on environmental factors are the design criteria for the operation of the digester. The design itself is determined by the engineering design needed to meet the criteria. In practice, the operational design factors that have the greatest bearing are those that pertain to duration of exposure of wastes to bacterial action, bacterial culture retention time, hydraulic detention time, and maintenance of proper temperature. The first three usually come under the heading of loading, since they determine loading rate, both solid and liquid.

Hydraulic detention time also is referred to as "digester replacement time" and usually expressed as "days detention" or "retention." Hydraulic detention time may be expressed as

$$t = \frac{V}{v}$$

where t = detention time, V = culture volume, and v = throughput per unit of time (i.e., same unit as t). For example, a 150,000-gallon ($568\ m^3$) digester operated at a daily throughput of 5,000 gallons ($19\ m^3$) would be on a 30-day detention period. Hydraulic loading depends not only upon type of system (i.e., high rate vs conventional, or combined), but also on the solids concentration of the feed. However, the latter generally is adjusted to suit the requirement of the digester culture.

Type of system exerts an effect not only through differentiation in rate of stabilization, but also through a difference in degree of separation of the liquid and solid phases. In the latter situation, it is possible to impose two distinct detention periods, one for the liquid phase and one for the solids. The detention periods with a high-rate system, not unexpectedly, are shorter than those with the conventional-rate systems. In practice, the second difference, namely, incapability of separating the solids from the liquid detention period, does not exist, since high-rate systems do incorporate a second step in which stratification takes place.

If the solids concentration of the feed is adjusted to provide the required solids retention time, the hydraulic detention time becomes synonymous with solids retention time. Ideally, the hydraulic retention time should be such that the suspended solids of the digester culture be within the range of 5 to 10 percent. Retention times applicable in the digestion of sewage sludge in general are also applicable to the digestion of other solids. Thus, for a typical conventional-rate digester, the retention time is from 30 to 60 days, whereas for high-rate digestion it is from 10 to 20 days. The designer should beware of a temptation to design on the basis of shorter-than-safe retention times. The temptation usually has its origin in extrapolations from laboratory studies involving conditions controlled at optimum to a degree impractical in a full-scale operation. This cautionary note should be kept in mind when using kinetic models, and their many variations to be found in the literature, to arrive at design numbers for retention periods.

An example of a kinetic model is one presented by Vesilund[91] in his text on the treatment of sludges. It describes a continuous steady-state completely mixed digester and is

$$\frac{(dM/dt)}{M} = (SRT)^{-1} = \left(\frac{a\mu_{max}S}{K_S + s}\right)^{-b},$$

where M = concentration of microorganisms (mass/volume); t = time; a = growth yield coefficient, time^{-1}; b = microorganisms decay coefficient, time^{-1}; μ = rate of waste utilization per unit volume, mass/volume/time; μ_{max} = maximum rate utilization per unit weight of microorganisms, time^{-1}; S = waste concentration, mass/volume; K_S = half velocity coefficient equal to the waste concentration when $\mu = \mu_{max}/2$; and SRT = solids retention time, defined as the ratio of total weight of active microbial solids in the system to the quantity of solids withdrawn daily. The constants are determined through laboratory solids analyses. A problem arises from the fact that the microbial population is estimated from a measurement of volatile suspended solids.

• *Loading* — Loading is determined by the rate of decomposition or stabilization and by which process is used, conventional or high-rate. The loading program followed traditionally with sewage sludge is on the order of 0.08 to 0.1 lb. volatile solids/cu. ft. culture volume/day (1.25 to 1.60 kg/m^3/day). Studies with wastes other than sewage sludge have shown that generally the maximum permissible loadings with such wastes exceed those with sewage sludge. Accordingly, loadings as much as 0.3 lb. refuse/cu. ft./day (4.81 kg/m^3/day) have been administered to acclimated digesters without a subsequent adverse reaction. In fact, cultures fed on wastes other than sewage sludge are far less sensitive to shock loads than those adapted to sewage sludge. The higher permissible loading with certain wastes may be due to a greater percentage of not easily decomposable material in those wastes.

■ **Structural:** The structural design of large-scale digesters is well advanced and is thoroughly documented in the literature. In this text, attention is directed mainly to volume and heat requirements. In Figure 17 are shown diagrammatic sketches which indicate the design arrangement for the three principal types of digestion systems.

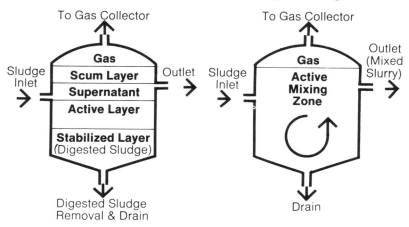

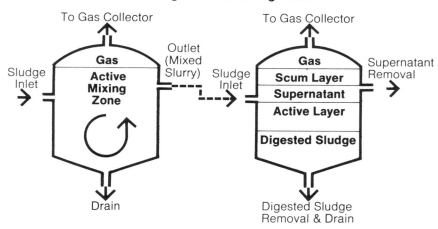

Figure 17: Design arrangements of the three different types of digestion systems

As stated earlier, the volume requirements are a function of the hydraulic detention period and feed slurry concentration. It can be expressed as

$$V = \frac{M}{C_p} \ t,$$

where p = density of the feedstream (lb./cu. ft. or kg/m³), V = reactor volume (cu. ft. or m³), M = refuse feed (lbs. or kg dry weight/day), C = feed concentration (%), and t = detention period (days).[12]

When mixing is a part of a process, it can be accomplished by one or a combination of the following means: (1) mechanical mixing, i.e., insertion of a stirring blade(s) into the culture; (2) recycling the slurry by withdrawing it from the bottom of the unit and discharging it at the top such that the recirculant is passed through the scum layer; and (3) recycling the gas. Gas recirculation of necessity differs from slurry recirculation in that gas is withdrawn from the plenum above the top of the culture and injected into the bottom of the culture. The rising and expanding gas bubbles accomplish the mixing. Each approach has its peculiar advantages and limitations. Probably all three functioning simultaneously would be needed when digesting refuse or manures that are mixed with bedding.

An important design consideration in anaerobic digestion is the provision of heat to maintain the digester contents at an optimum temperature. The heat itself is supplied either by way of a heat exchanger (pipe "coil") placed in the digester and, where mixing is accomplished by slurry recirculation, by heating the recirculant. The greatest expenditure of heat energy is that required to raise the temperature of the incoming feed to the level required for the culture. The heat required (heat input) to do this is proportional to the mass flow rate and the difference between the feed stream and the reactor temperatures. It may be expressed as

$$HI = SC(T_1 - T_0),$$

where S = the feedstream (lbs./hour or kg/hour), C = the specific heat content (Btu/lb. -°F or Cal/kg - °C); T_1 = reactor temperature (°F or °C); T_0 = feedstream temperature (°F or °C); and HI = heat required (Btu/hour or Cal/hour). (Btu/hour can be converted to Cal/hour by multiplying by the factor 252.)

Other factors leading to an energy expenditure for heating are the need to compensate for the heat lost from the digester by convection and radiation to the ambient atmosphere and by way of evaporation of the water vapor in the gas stream. However, studies have shown that the energy involved in such heat loss is minor in comparison with the expenditure needed to heat the feedstream where a large-scale digester is concerned.[139] The heat retention capacity of a full-scale digester is prodigious.

■ **Small-Scale Digesters:** A considerable amount of literature on small-scale digestion exists in the form of pamphlets, brochures, and papers. Among them are two from the New Alchemy Institute;[140,141] two papers and a pamphlet written by Ram Bux Singh;[125,142,143] and another, as a section of a WHO monograph on composting, by Gotaas.[34] Unfortunately, it may be difficult to obtain a copy of Singh's pamphlet, and the monograph by Gotaas is out of print. However, an excellent summary of Singh's design concepts as applied to the digestion of animal manures, especially dairy manure, is contained in a report prepared by the Ecotope Group for the state of Washington Department of Ecology[144] and in papers by Singh in *Compost Science*.[142,143] Incidentally, the Ecotope report can serve as an excellent model for determining the feasibility of a digestion operation for a specific situation.

The structure needed for small-scale digestion operations, i.e., for farms and other producers of putrescible organic wastes, obviously are less complex than those for municipal operations. Nevertheless, when the scale of the operations goes beyond a few cubic feet (15 to 20 cu. ft. or 0.5 to 0.75 m^3), the design requirements become more complex than for a compost operation involving the same amount of material. Singh classifies a digester as small if it produces less than 500 cu. ft. (14.15 m^3) gas per day.

Before describing specific digester designs, it should be pointed out that the materials used in constructing all but the smallest of digester systems should be durable, strong, and resistant to corrosion. Moreover, they should be of such a nature that the digester is gas- and water-tight. The digester unit itself can be made of any type of blocks capable of being fashioned to contain liquid. Sinking the digester in the earth may compensate for deficiencies in the structural strength of the digester walls. In fact, where it is feasible, the digester tank should be buried to within the top 1 or 2 ft. (30 to 60 cm). The reason here is not so much for strength as for insulation. An idea of the materials involved can be gained

TABLE 11.   MATERIALS REQUIRED FOR A DIGESTER TO
PRODUCE 100 CU. FT. (2.8 m^3) GAS PER DAY[135]

Material	Quantity
Cement	40 bags
Sand	300 cu. ft. (8.5 m^3)
Brick ballast	100 cu. ft. (2.84 m^3)
Bricks	7500
12 or 14 gauge M.S. sheet drum, 5 ft. (1.5 m) in diameter and 4 ft. (1.2 m) in height. Open at bottom.	
M.S. angle iron for structure and gas holder guide	100 ft. (30 m)
Alkaline pipe 0.5 in. (0.125 cm) in diameter	50 ft. (15 m)
Alkaline pipe fittings B end, elbow and sockets of 1 in. (2.5 cm) and 0.5 in. (1.25 cm) fittings	3 each
Wire gauge — 80 mesh	1 sq. ft. (0.093 m^2)
Misc. fittings	
Paint (enamel)	1 gallon (3.8 liters)

from a list compiled by Singh[143] for a 100 cu. ft. (2.8 m^3)/day gas production. The list is repeated in Table 11.

Any one of a number of structural designs may be selected. The choice should be made to fit the needs of the particular situation or application. One of the drawbacks of digestion is the need for a relatively complex structure if a significant amount of wastes is involved. Attempts to curtail the fulfillment of these needs lead to operational failures and safety hazards. One such hazard would be an inadequate provision for storage of the gas, a constituent of which is methane (CH_4). Methane is explosive at concentrations as low as 5 percent in air.

The required volume of the digester determines to a large extent the design of the plant, especially with respect to complexity. Volume can be determined on the basis of loading rate in the manner described in the section on large-scale digestion. Where it is difficult or infeasible to determine dry weight and volatile solids, a rule of thumb estimate can be made on the basis of 1 cu. ft. volume/35 lbs. dung (1 m^3/53 kg).[142] A somewhat larger volume should be allowed for digesting vegetable wastes because of their relatively low density and of the large volume of water needed to slurry them.

The formula given in the preceding section and the data listed in Table 12 can be used in designing a small scale-digester both with respect to volume of the digester and to the amount of gas production to be expected. In Figure 18 is shown a schematic

TABLE 12. MANURE PRODUCTION BY VARIOUS
TYPES OF ANIMALS

Animal	Volatile Solids of Feces in %	
Dairy cow (1,400 lbs., 635 kg)	12.1	5.5
Heifer (900 lbs., 408 kg)	7.1	3.2
Horse (850 lbs., 386 kg)	5.5	2.5
Swine (160 lbs., 73 kg)	1.3	0.6
Sheep (67 lbs., 30 kg)	0.4	0.18
Chickens (3.5 lbs., 1.6 kg)	0.06	0.03
Human		
feces	0.13	0.06
urine	0.10	0.045

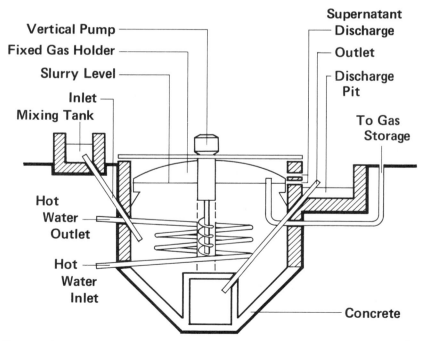

Figure 18: Digester design for the production of 2500 cu. ft. (70.75 m³) biogas/day in cold climates

arrangement of a two-stage digester as proposed by Singh.[143] It is a good representation of the relation of the various digester components to each other and indicates methods of feeding and gas collection.

A detailed plan of a single-stage digester is shown in Figure 19.[34] With this type of digester (fixed-cover), provision must be made for collecting the gas formed in the digester. Accordingly, the gas is piped to a gas collector. As a matter of safety a fixed-cover digester should always be completely filled with slurry. The reason is the difficulty of completely excluding air from the unit. If any air should leak into an incompletely filled unit and through some mischance an open flame or spark occurs, the results could be disastrous.

A cheaper approach to digester design would be the use of "off-the-shelf" manure storage tanks.[144] A gas holder may consist of an inverted tank telescoped into another which is filled with water. The inverted tank functions as a floating cover and moves up and down through the water as determined by the quantity of gas discharged into it from the digester(s). The water in the annular space between the tank wall serves as an effective seal against the escape of gas. The cover should be provided with suitable guides for its vertical movements. If not already heavy enough, it should be weighted (or, if too heavy, counterweight it) such that a line-gas pressure of 4 to 8 in. water (10 to 20 grams/cm^2) be provided.

A digester can be designed such that gas can be safely collected in the digester unit itself. This can be done by designing the unit itself to have a floating cover as shown in Figure 20. As with a separate gas collector, the water in the annular space in which the vertical walls of the cover are positioned acts as a gas seal. As gas is generated, it lifts the cover. The cover usually is made of 2- to 3-mm-thick sheet iron strengthened and framed with angle iron or cross braces. The vertical movement of the holder is guided by a system of rollers and U-shaped guides fastened to the cover. Again, as with a separate gas collector, the cover should be heavy enough to provide the necessary line-gas pressure. If it is too heavy, it should be provided with a counterweight.

Gotaas' monograph[34] describes a system for handling human excreta as well as manures. The plans suggested by him are sketched in Figure 21. A suitable water trap should be placed

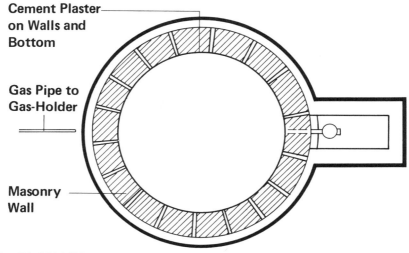

Cement Plaster on Walls and Bottom

Gas Pipe to Gas-Holder

Masonry Wall

A. PLAN VIEW

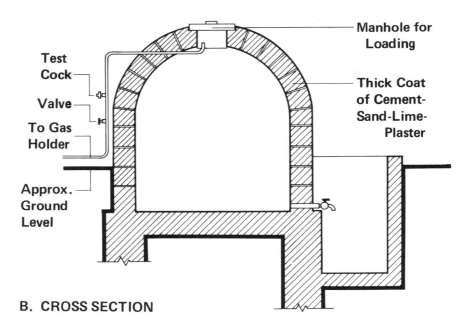

Manhole for Loading

Test Cock

Valve

To Gas Holder

Thick Coat of Cement-Sand-Lime-Plaster

Approx. Ground Level

B. CROSS SECTION

Figure 19: Design of a small, farm-scale digester[33]

Gas Storage Space Below Floating Cover

Floating Cover

Water Seal

Initial Water Level

Gas Pipe

To Fixtures

Reinforced-Concrete Walls (masonry walls to be thicker)

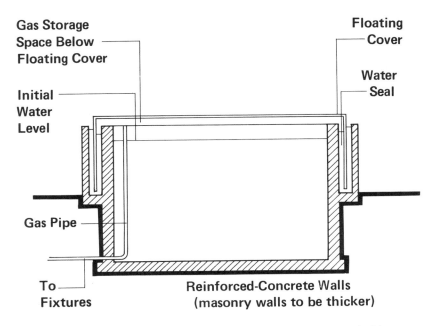

Figure 20: Digester with floating cover serving as the gas holder

between the digestion tank and the latrine pit to prevent the escape of gas. Moreover, the water in the latrine should be at a level high enough to ensure a clean and sanitary operation.

Other design approaches for small-scale digesters are being proposed, but at the time of this writing none has been tested on a meaningful scale. A favorite alternative among the proposals is to set up a horizontal unit instead of the conventional vertical type. Theoretically, a horizontal unit would be expected to be readily adaptable to a separation of the acid and methane phases, which supposedly would result in an increase in digester efficiency. Hopefully, the separation could be easily accomplished with a horizontal unit because the flow of the wastes through the unit would match the sequence of digestion. The reasoning is that since sequentially the acid phase is the first phase, it would take place at or near the point of insertion of the wastes into the horizontal digester. As the solids move through the unit, the acids gradually are transformed into gas and low-molecular-weight decomposition products. Eventually a zone is reached at the distal end of the unit

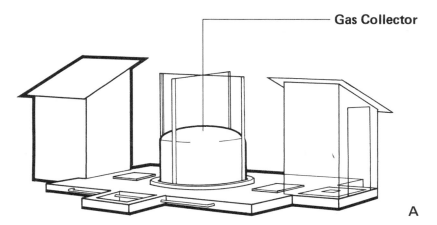

Figure 21(**A**): Latrines and manure biogas plant complex—arrangement of latrines and digester[33]

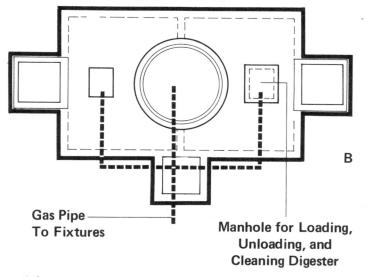

Figure 21(**B**): Latrines and manure biogas plant complex—plan of latrines and digester[33]

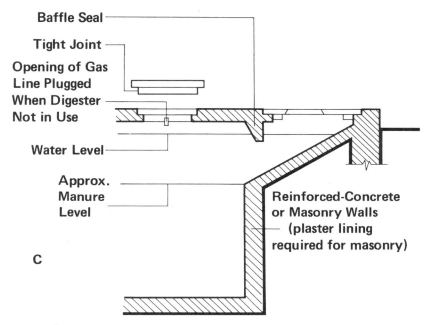

Figure 21(**C**): Latrines and manure biogas plant complex—cross section of latrine and digester[33]

where the methane production step is the predominant one. In addition to increased efficiency, it is theorized that a unit of such a design would be less demanding in terms of structural strength. It has even been suggested that a rubber container much like that used in trucking certain liquids could be used. Unfortunately, experience thus far with horizontal digesters has been attended by a singular lack of success.

RESIDUES AND THEIR DISPOSAL

The two principal residues are the supernatant and the settleable (suspended) solids. The two can be disposed of in a combined form (i.e., the digested slurry) or individually. For the components to be handled individually, the digester slurry must be allowed to remain quiescent long enough for the suspended solids to settle out of suspension. This is accomplished in the second digester of a two-stage process and partially in a conventional-rate system.

Supernatant

The nature of the supernatant reflects that of the waste being handled. It contains the soluble nutrients in the digester culture, e.g., nitrogen, phosphorus, potassium, etc. This soluble portion may include the greater part of the nitrogen content of the slurry, especially if manures are digested. Therefore, it should be put to a productive use, which ultimately is application on the land to enhance soil fertility. In so doing, the constraints described for the land disposal of sewage sludge should be observed. For operations involving the digestion of refuse, the supernatant can be used as makeup water to slurry the refuse, especially when sewage sludge is not digested in combination with the refuse. This usage would apply to any waste not already in a slurry form.

Settleable Solids

As stated previously, the elemental makeup of a sludge is largely determined by that of the raw material fed the digester. Thus, sludges from the digestion of raw wastes rich in nitrogen have a percentage of nitrogen greater than that in the sludges from nitrogen-poor wastes. (The noncellular nitrogen is mainly in the reduced form (NH_4-N); whereas in compost it usually is in the oxidized form (NO_3-N)). The same holds true for the other elements. Generally, therefore, the sludge from the digestion of refuse has a lower nutrient content, especially of nitrogen, than do sludges from manures. On the basis of dry weight, the nitrogen content of digested hog manure sludge ranges from 6.1 to 9.1 percent; of digested chicken manure, 5.3 to 9.0 percent; digested cow manure, 2.7 to 4.9 percent; and digested refuse, 0.6 to 2.5 percent.

The physical characteristics of sludge from refuse digestion differ from those of digested agricultural wastes in that it may have larger particles and certainly a greater proportion of nonbiodegradable debris than found in digested manure. The exception would be sludge from the highly segregated organic refuse produced in the Cal Resource Recovery system,[145] because in that system all nonbiodegradable materials are sorted and recovered.

If the principal motivation or objective of digesting the wastes is energy production, then possibly incineration should be the approach to follow in the disposal of the sludges from refuse digestion. The reason is that a significant portion of the potential energy content of the refuse is to be found in the digested residue. This is true regardless of whether the residue is from a single exposure to the digestion process or from a second exposure following the special conditioning treatment described earlier.[137] According to Pfeffer,[145] recovery of energy solely as CH_4 through the digestion of refuse results in a recovery efficiency of only 32.6 percent as contrasted to 67.4 percent if the energy from incinerating the sludge is included. However, incineration necessitates dewatering the sludge, a process that can be expensive in terms of money and energy. Added to the dewatering costs are those of preventing air pollution due to the incineration.

If the sludge is not incinerated, it can be landfilled or spread upon the land. Except in the unlikely event of the presence of a toxic substance in the sludge, the second alternative would seem to be the one of choice.

EVALUATION OF THE PROCESS

Again it should be emphasized that this chapter on anaerobic digestion does not deal with the digestion of sewage sludge. Therefore, the present section on evaluation pertains only to the anaerobic digestion of nonsewage solids. The evaluation is considered from two aspects: anaerobic digestion as a waste-treatment process and anaerobic digestion as a method of recovering (tapping) the energy in solid wastes.

Digestion as a Treatment Method

The evaluation of digestion as a treatment method is largely based on costs inasmuch as they have the decisive influence on feasibility.

■ **Refuse:**   The extent of total and volatile solids destruction of the biologically degradable fraction of solid wastes matches that accomplished in composting, namely, from 20 to 40 percent of the total solids and 30 to 60 percent of the volatile solids. The extent of the destruction of the organic residue from the Cal Resource Recovery process is on the order of 70 to 80 percent.[139] Destruction of the cellulosic fraction can be as high as 80 percent.[12,131] Naturally, the total amount of destruction depends upon the degree of segregation of the various constituents of the refuse, operating conditions, and especially on length of detention period. An example of the effect of temperature and detention time is illustrated by the relation of the slopes of curves in Figure 22, which are based upon data reported by Pfeffer.[12]

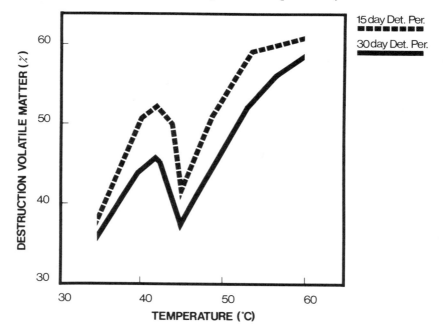

Figure 22:   Effect of temperature of digestion on percentage volatile solids destruction

At the time of this writing, the reported costs for anaerobic digestion of refuse were based upon the results of small, pilot-scale experimental runs, on computer extrapolations of those results, and on experience with sewage-sludge digestion. This should be kept in mind when evaluating the various cost analyses made at the time. As of 1975, costs for digestion were comparable to those for incineration in terms of processing and with no account made for sale of the energy. The estimated capital costs per ton of capacity were on the order of $9,200 to $14,000/ton ($10,141 to $15,430/metric ton).[145,146] Operating costs exclusive of gas processing ranged from about $4.00 to $8.00 ($4.40 to $8.80/metric ton) of wastes processed. The costs include those for shredding, sorting, and the actual digestion.

The higher numbers also cover those for incinerating the digestion residue. If incineration were included, of all of the components it would constitute the largest individual item, i.e., almost 35 percent of the total capital cost. With regard to preparation costs, it should be pointed out that, as in composting, they would be shared with other recovery steps in a full resource recovery operation.[147]

■ **Agricultural Wastes:** Studies on the anaerobic digestion of agricultural wastes have been almost exclusively concerned with the disposal of animal manures and cannery wastes. However, until the last few years, when energy became a critical factor in agriculture in the United States, anaerobic digestion was not held in high regard, and aerobic or facultative systems were preferred as means of treating such wastes. In fact, a perusal of the proceedings of two widely attended conferences on animal waste management[148,149] revealed only one paper solely concerned with anaerobic digestion. In two others, except for anaerobic lagooning, it was mentioned as a possibility and then dismissed. The absence of interest stemmed from the high cost of the capital equipment as well as from the existence of certain process problems. For example, if manures have a high nitrogen content, e.g., swine manure, an excessive amount of ammonia accumulates in the culture, with a consequent inhibition of digester activity due to the toxicity of the ammonium ion.[150] The buildup can be minimized by a liberal dilution of the wastes with water and with highly carbonaceous wastes. Reported volatile solids destruction values range from 30 to 60 percent of the incoming solids.

Reports on costs of anaerobic digestion of agricultural wastes are as scarce as those on the process itself. One of the earlier articles[151] quotes an initial cost for a digester capable of digesting the wastes from 100 dairy cattle and 100 hogs as ranging from $9,000 to $15,000 (1963 dollars — or about $23,000 to $38,000 in 1975 dollars). A more recent analysis[144] projects a cost of about $56,700 (1974 dollars) for a complete plant capable of processing the wastes from 350 cattle units (1 unit equals 1,000 pounds of animal). A breakdown of the costs is presented in Table 13. The effect of scale-up on such a system is indicated by the data in Table 14.

TABLE 13. ESTIMATED COST FOR BIOGAS PLANT BASED ON 350 DAIRY-CATTLE UNITS[144]

Direct Cost	Labor	Material
Two tanks	$ 300	$13,500
Conditioning tank	100	1,000
Effluent tank	100	1,000
Surplus heater and pump	50	1,500
Enclosure structure	150	2,000
Piping for digester	25	600
Piping for gas	50	1,500
Input pump	25	3,000
Exit pump	25	3,000
Insulation	150	2,750
Gas recirculation mixer	100	2,000
Gas compressor	25	1,500
Gas storage	50	4,500
Safety equipment	150	2,000
Generator and gas engine	100	6,000
Subtotal	$1,400	$45,850
		+ 1,400
		47,250
Add 20% contingencies =		+ 9,450
Total direct cost =		$56,700

TABLE 14. EFFECT OF SCALE-UP ON COST OF BIOGAS
PLANT UTILIZING DAIRY-CATTLE UNITS[136]

Digester Volume		Cattle Units	Multiplier	Monthly Cost ($)	Cost per Cattle Unit[a] ($)
(cu. ft.)	(meters³)				
50,000	1415	175	———	318.00	1.82
100,000	2830	350	1.0000	400.76	1.15
150,000	4245	525	1.2754	511.14	0.97
200,000	5660	700	1.5260	611.56	0.87
250,000	7075	875	1.7488	700.85	0.80
300,000	8490	1050	1.9545	783.29	0.75

[a] One cattle unit = 1,000 lb. (453.5 kg) of animal.

The monetary returns from the unit can be estimated from the calculated value of the fuel gas and fertilizer (sludge) produced by the system. The estimated yearly net production of biogas is 1,875,829,230 Btu (198 × 10⁹ kj). The estimated production of ammoniated nitrogen in the form of sludge would be about 8 tons/year (7.2 metric tons/year).

Cost analyses were made by Schmid[152] for the production of gas from the digestion of the manure from a 35,000-head feedlot under three different sets of conditions. In one set, a concrete open feedlot was assumed, wastes would be scraped out only when the cattle were loaded-out at about 150-day intervals, and the wastes would be stockpiled for feeding to the digester. In a second set, a total confinement system was assumed in which all manure and urine would be trapped and transported as a slurry to the digester. The assumed waste temperature of the slurry at the time of introduction into the digester was 70°F (21°C). A third set involving a dirt open feedlot also was assumed. However, since the last named setup resulted in a negative recovery of energy, no cost analysis was made for it.

Based upon a 15-day detention period, the digester capacity needed for each set of conditions was estimated to be 500,000 cu. ft. (14,000 m³). The capital cost for the digester was calculated to be $1,000,000 (1975 dollars); amortization of capital costs,

$290/day; and operation and maintenance, $110/day. Cost of the gas to be produced would be $2.65/1,000 cu. ft. ($0.094/m^3) of methane. The second set of conditions (closed confinement) would entail a capital cost of $1,500,000 for a 750,000 cu. ft. (21,240 m^3) digester. Amortization and capital costs would come to $435/day and operation and maintenance, $165/day. The cost of the gas produced would be $0.63/1,000 cu. ft. ($0.022/m^3). Schmid emphasizes that in making the last-named cost analysis, no costs were factored in it for the greatly increased cost of confinement feeding and for other problems arising from that mode of feeding. The main reason for resorting to confinement feeding would be the existence of governmental regulations of such a nature as to leave confinement feeding as the only course open to the feedlot operator. Under that circumstance, anaerobic digestion would be a means of recouping at least a part of the higher cost of establishing and maintaining a confinement feedlot.

Digestion as an Energy Source

Depending upon the convictions of the writer, a good case can be made for or against the practicality of anaerobic digestion as a means of recovering the energy contained in wastes. This is especially true with reports concerning the anaerobic digestion of refuse and much less so with respect to animal manures from operations having a high animal density per unit of area. The flexibility stems from the extent of the scope of the energy balance chosen by a particular writer. The scope can range from a complete balance down to the very limited one termed "processing balance" in this text for want of a better term.

A "processing balance" covers only those energy expenditures which begin with the delivery of the wastes to the plant and are operational in origin. To make a processing balance, one includes the energy content of the wastes to be digested, the energy expended in accomplishing the necessary pretreatment, the operation and heating of the digester, the disposal of the sludge, the preparation, and the total and the net energy yield from the process.

In making a complete balance, not only are the processing energy expenditures and the energy content of the wastes taken into consideration, but also the energy involved in the production of the components of the system. An example is the total energy expended in quarrying and processing the limestone used in fabricating the digester, as well as that consumed in all the intermediate steps between the manufacture of the concrete and the actual utilization of the material to build the digester. This input would apply to all other materials used in constructing the digester. Finally, the energy expended in constructing the digester would be added, as well as that in fabricating the other components of the system.

The calculation of such a complete energy input obviously would be a tedious task and certainly would be attended with a significant degree of uncertainty. Yet, logically it is the only approach that would make possible a true comparison between competing systems for energy recovery. The reason is that the process energy input for one system may be lower than that for a competing system; whereas the complete energy input might be much greater. This is especially true where complex structural components are required, as is the case in anaerobic digestion and incineration. The problem is solved in everyday practice by relying upon comparative economics (i.e., costs) of processes. Indeed, costs probably are a reliable indicator of comparative efficiencies because, as many economists maintain, in this day and age costs ultimately are based upon energy.

Another source of flexibility is the lack of adequate data on the energy requirements of the separate steps in the process, e.g., size reduction of the wastes, slurrying the wastes, mixing the digester contents, etc. When data are available, the range of values for a given step are very wide. The consequence is that the choice of a number unfortunately often reflects one's enthusiasm about the process, albeit subconsciously.

For making a complete balance, information regarding energy expenditures for the various steps leading up to the anaerobic digestion step itself, as well as for all the items in the physical structures, is available in an assortment of handbooks and statistical abstracts. The listing of this voluminous collection of information is beyond the scope of this book. Instead, data pertinent to arriving at an operational balance are presented herein.

Under the category of available energy is that in the raw material to be digested. The reported calorific values of refuse range from 6,000 to 7,500 Btu's/lb. dry solids (1.39 to 1.74×10^4 kj/kg). The calorific content of agricultural wastes averages about 8,000 Btu/lb. (3,840 kj/kg), i.e., average value for dry organic matter in general.

The energy required for processing refuse for digestion, heating and mixing the digester contents, pumping, etc., is greater than that for manures. On the basis of each ton of refuse received and processed in a 1,000-ton (907 metric-ton)/day plant, Pfeffer[145] estimates that a total power requirement would be on the order of 602,045 Btu/ton (700,283 kj/metric ton) processed (2,780 hp, 30 percent efficiency); and energy to heat the digester to 60°C, about 1.08×10^6 Btu/ton (1.25×10^6 kj/metric ton) refuse at 85 percent efficiency. Thus, according to his estimate, the total input of energy after deducting that from incinerating the sludge would be about 1.68×10^6 Btu/ton (1.95×10^6 kj/metric ton).

The energy produced by the system would be in the form of methane and, if the sludge were burned, of steam. Reports on the yield of methane from digesting refuse range from 2 to 5 cu. ft./lb. (125 liters to 312 liters/kg) dry organic solids introduced into the digester culture. For a 1,000-ton (907 metric ton)/day plant, according to Pfeffer's extrapolations, the expected CH_4 yield per ton of refuse received and processed would be on the order of 3,000 to 3,600 cu. ft./ton (83.7 to 112.4 m^3/metric ton) or, in terms of energy, from 2.9×10^6 to 3.50×10^6 kj/metric ton of refuse processed.

Since a large part of the input energy remains locked in the residue (sludge), Pfeffer recommends that it be recovered by way of incineration accompanied by steam production. The heat energy thus recovered would amount to 3.305×10^6 Btu/ton ($3,853 \times 10^6$ kj/metric ton) refuse received and processed in the plant. The energy recovery of Pfeffer's hypothetical 1,000-ton/day plant would be about 32.6 percent if methane only were produced. If the sludge were incinerated and a market existed for the steam thus produced, the recovery efficiency would advance to 63.4 percent.

In making an energy balance for agricultural wastes (crop residues and manures), a calorific value of about 8,000 Btu/lb. (18,568 kj/kg) dry organic solids can be assumed. In general, methane production would be on the order of 4 to 5 scf/lb. (290 liters to 312 liters/kg) introduced into the digester. Using the energy consumption values named by Pfeffer for preparation of the wastes and operating the digester, an average energy recovery of about 35 to 50 percent could be expected from the digestion of manure and crop residues as delivered to the plant.

There are various ways of arriving at specific estimates of gas production to be expected from the digestion of manure. According to Singh's publication,[125] from 3.1 to 4.7 cu. ft. of gas are produced per pound (0.20 to 0.29 m^3/kg) of total solids in the manure. Using the number of cattle (dairy) as a basis, then from 35 to 52 cu. ft. (0.9 to 1.5 m^3) of gas can be expected per cattle unit (i.e., per 1,000 lbs. or 453.6 kg of animal weight). About 7 cu. ft. of gas can be obtained from each pound (0.44 m^3/kg) of vegetables digested. Gas production also can be estimated on the basis of C.O.D. In the digestion of sewage sludge, about 5.62 cu. ft. are generated per pound (0.35 m^3/kg) of manure introduced. The C.O.D. of the manure from each cattle unit is about 9 lb. (4.1 kg). Therefore, on the basis of C.O.D., the gas production per cattle unit would be 9.0 $\times$ 4.5 $\times$ 0.8 or 32.4 cu. ft. (0.92 m^3). Amounts of gas actually produced from the digestion of various wastes are listed in Table 10. (An idea of the amount of gas needed for certain useful applications on the farm may be gained from the data in Table 15.)

For estimating the energy consumption needed to collect and transport crop residues, numbers for energy requirements in the form of fuel (gasoline) are available. Collecting crop residues at the field would use about 0.45 percent of the gross energy of the residue harvested. The estimate is based upon a consumption of 0.6 gallon (2.27 liters) gasoline by a tractor to collect a ton of dry organic matter at 2 tons/acre (4.48 metric tons/ha). Hauling the residue to a transfer station (2 miles or 3.21 km from the field) would require about 0.11 gallon (0.4 liters) gasoline/ton (i.e., the equivalent of 13,000 Btu/acre or 33,890 kj/ha), or about 0.8 percent of the energy content of the material. The amount of

TABLE 15.   GAS REQUIREMENTS FOR VARIOUS
DOMESTIC APPLIANCES[142]

Application	Specifications[a]	Gas Consumption (cu. ft./hr.[b])
Gas cooking	2-in. diam. burner	11.5
	4-in. diam. burner	16.5
	6-in. diam. burner	22.5
Gas lighting	1-mantle lamp	2.5–3.0
	2-mantle lamp	5.0
	3-mantle lamp	6.0
Refrigerator	18 in. X 18 in. X 18 in. (flame operated)	2.5
Incubator	18 in. X 18 in. X 18 in.	1.5–2.0
Boiling water		10 cu. ft./gal.
Running engines	(converted diesel or gasoline)	16–18 cu. ft./hp/hr.

[a]To convert inches to centimeters (cm), multiply by 2.54.
[b]To convert cubic feet (cu. ft.) to meters, multiply by 0.0283.

energy consumed at the transfer station would be minute. Transporting the residue from the station to the digester would require about 2,800 to 3,500 Btu/ton-mile or 1,953 to 2,442 kj/metric ton-km.

In describing Schmid's cost analysis of energy recovery from feedlot wastes, it was pointed out that the net energy production from the digestion of the manure from a predominantly dirt feedlot would be negative. A reason for the negative energy balance is the biological destruction of a large part of the biodegradable organic matter that took place during the 150-day periods between cleaning the lot, as well as during the storage of the wastes during the time they were being fed to the digester. The loss is worsened by that of the nitrogen from the urine which soaks directly into the ground. On the other hand, with a concrete open feedlot, he showed a net gas production of 151,000 cu. ft. (4,276 m^3), but at a cost per unit of gas from three to five times that of the retail price of natural gas in 1975.

The energy balance was the most favorable with a total confinement system. The net gas production with that setup was

972,500 cu. ft. (27,543 m^3) produced at a cost only slightly above that of the present retail price of natural gas. (This amount of gas would be enough to heat nearly 1,000 homes in Kansas.) The reason for the favorable showing is the minimization of loss of organic matter through biodegradation prior to introduction into the digester because of intervening storage periods. Moreover, the wastes do not have time to cool off, and hence less energy is required to raise the feed temperature to that of the digester culture.

The data used by Schmid in making his energy analyses and his resulting energy balances are presented in Table 16.

TABLE 16. ENERGY BALANCES FOR GAS PRODUCTION
FROM THE DIGESTION OF FEEDLOT WASTES FROM
35,000 HEAD OF CATTLE[152]

	Type of Feedlot[a]					
	Dirt		Concrete		Confined	
	English Units	Metric Units	English Units	Metric Units	English Units	Metric Units
Digester Capacity: 1. total dry solids (lbs. & kg/day)	350,000[b]	158,750	208,000	94,349	315,000	14,288
2. degradable organics (lbs. & kg/day)	28,000	12,700	42,000	19,051	147,000	5,668
Organic Loading (lbs. or kg degradable solids) (1,000 cu. ft. or 280 m^3 of digester capacity)	33	15	86	39	194	88
Gas Produced (100% CH$_4$) (cu. ft. or m^3/day)	224,000	6,339	336,000	9,508	1,176,000	33,281
Gas required/day (cu. ft. or m^3): 1. heating to 95°F (35°C)	173,000	4,896	110,000	3,113	92,500	2,618
2. digester heat loss	123,000	3,481	74,000	2,094	110,000	3,113
3. total per day (cu. ft. or m^3)	296,000	8,376	185,000	5,236	203,500	5,759
Net gas production/day (cu. ft. or m^3)	72,000	20,371	151,000	4,273	972,500	27,522

[a] See text for detailed description.
[b] Allows for a sizeable amount of soil scraped up with the manure

To temper the reader's optimistic conclusions on the potential of digestion in the recovery of energy from wastes, it should be again pointed out that the numbers refer only to the operational aspects of the process, and do not take into consideration the "background" energy requirements. That these latter are substantial is attested by the high capital costs. Until relatively recently the latter have militated against an extensive utilization of anaerobic digestion in the treatment of agricultural wastes. It is only when energy sources became in short supply that interest arose in the utilization of the process. As far as refuse is concerned, the case is yet to be proven in favor of digestion as a competitive means for recovering the energy tied up in them.

4. CONVERSION OF ORGANIC WASTES INTO YEAST AND ETHANOL

INTRODUCTION

A considerable amount of attention has been directed toward the utilization of cellulose, a component which constitutes a considerable fraction of organic solid wastes both urban and agricultural in origin. As a result, several approaches have been proposed and adapted for accomplishing the utilization. Among the more conspicuous are the use of the material as a fuel to generate heat energy or to recycle the material as a raw material for paper manufacture.

One of the less tried approaches is the conversion of cellulosic carbohydrates in wood and paper or paper products to a single-cell proteinaceous feedstuff or to ethanol.[143,154] The approach is based upon the concept of hydrolyzing the cellulose carbohydrates to sugars and then using the sugars as a substrate in the production of yeasts. The yeasts can be utilized directly as a feedstuff or can serve as the active agents in the fermentation of the sugars to ethanol. The hydrolysis can be accomplished by chemical processes or through enzymatic reactions (i.e., biologically). In another approach, the hydrolysis step is by-passed, and the yeast (feedstuff) is cultured directly on the cellulose. This latter involves the exposure of the wastes to a strong alkaline solution to alter their physical characteristics and those of the cellulose such that the permeability of the latter is enhanced. This method has been explored as a means of converting bagasse into single-cell protein.[155,156]

Because of the high capital and operational costs involved, the idea of converting cellulose to ethanol or to a feedstuff had little appeal in solid-waste treatment during the national period of plenty (1950s to 1960s). A strongly dissuasive factor was the fact that the only possibility of making the approach even remotely economically competitive with other more conventional methods would be the sale of the product of the process, namely sugars, single-cell protein, or alcohol. Unfortunately for the process, during the "period of plenty" those items were in abundant supply and were available at quite low prices. Now, the situation has changed drastically and a period of chronic shortages has set in. As a consequence, the monetary value of the three products has increased to a level at which the economics of the conversion of cellulose to a sugar feedstuff or to an energy source begins to appear to be more favorable.

HYDROLYSIS

Since a discussion on hydrolysis is more readily understood if it is preceded by a description of the chemistry of cellulose, a brief description of the latter is given here. Cellulose is essentially a polymeric material consisting of anhydrous residues. The glucose units are in glucopyranose form and are linked together by an ether bridge, connecting position 1 of each molecule to position 4 of the adjacent one. The structure is represented as follows:

The degree of polymerization of natural forms of cellulose is typically 3,000. Cellulose molecules occur in straight chains. Hydrogen bonding exists between parallel molecules in some regions of natural cellulose, a characteristic which imparts a crystalline nature. The noncrystalline regions are termed *amorphous*. The possession of a crystalline structure increases the resistance of cellulose to enzymatic reaction and subsequent hydrolysis but not necessarily to acid hydrolysis, since the acid can diffuse into the crystalline regions.

Chemical (Acid) Hydrolysis

■ **Chemistry and Kinetics:** Acid hydrolysis generally involves the subjecting of the selected waste to a treatment, the essential elements of which are an acidified suspension, a medium (water), and elevated temperature and pressure. If wood is hydrolyzed, the reaction is

wood cellulose → sugars → decomposition products.

(In practice, the reaction is stopped at the sugar stage.) Wood consists of carbohydrates (70 to 80 percent) and lignins (20 to 30 percent). The lignin does not hydrolyze under the hydrolysis conditions applied in practice. The carbohydrate fraction consists mainly of alphacellulose (40 to 50 percent) and hemicellulose (20 to 30 percent). Alphacellulose (true cellulose) is hydrolyzed into its glucose units:

$$(C_6H_{10}O_5)_n \ + \ H_2O \ \xrightarrow{\text{acid}} \ nC_6H_{12}O_6$$
alphacellulose

The hydrolysis of the hemicellulose results in the production of glucose as well as pentoses, of which xylose is the principal one. The total hydrolysate of wood generally is from 60 to 68 percent glucose, 2 to 25 percent mannose, 0 to 6 percent galactose, 8 to 30 percent xylose, and 0 to 3 percent arabinose. The percentages are a function of the type of wood. Paper consists principally of true cellulose, and therefore the primary sugar from it is glucose. The yield of potential reducing sugars ranges from 64 to 75 percent with the hardwoods and from 52 to 70 percent with the softwoods. The respective ranges for yield of potential fermentable sugars are 40 to 57 percent and 40 to 60 percent.

The key factors that affect the rate of hydrolysis are particle size (surface area to mass ratio), liquid to solids ratio (L/S), acid concentration, temperature, and time. The hydrolysis rate increases with increasing L/S ratio. However, in practice an upper limit is placed upon the ratio by way of a trade-off between gain in rate and increase in cost of equipment and operation due to the larger volumes to be processed at the higher L/S ratios. The trade-off apparently comes at an L/S of 10/1. The yield of sugar also increases with increase in acid concentration and rise in temperature. The limiting factor on concentration of acid is corrosion.

Moreover, unless the acid can be recovered, the use of high concentrations results in an increase in raw material costs. However, recovery is difficult to accomplish. A maximum concentration of 0.5 percent (H_2SO_4) seems to be the optimum in terms of costs and yet enough to ensure rapid hydrolysis. It is economically more feasible to compensate for high acid concentration and L/S ratio by elevating the temperature. A 200°C level seems to be a satisfactory one.

Saeman[155] showed that the hydrolysis of the crystalline wood cellulose is an A $\xrightarrow{K_1}$ B $\xrightarrow{K_2}$ C reaction as determined by the equations

(1)
$$\frac{dC_A}{dt} = K_1 C_A$$

(2)
$$\frac{dC_B}{dt} = -K_2 C_B K_1 C_A$$

where C_A = cellulose concentration, C_B = sugar concentration, K_1 = reaction rate constant for cellulose to sugar, K_2 = reaction rate constant for sugar to decomposition products, and t = time,

$$K_t = P_i e - \Delta H/RT \quad i \ 1,2$$

Fagan *et al.*[156] demonstrated that the hydrolysis of cellulose in paper effectively followed Saeman's model for wood cellulose with the exception of a difference in rate constants, as would be expected. As a result of experimentation, they found that the reaction rate could be represented by

(a) Decomposition of cellulose (A $\longrightarrow$ B)

$$r_1 = 28 \times 10^{19} C^{1.78} \ \exp(-45,100/RT)C_A ,$$

where C = acid concentration in weight percent and C_A = cellulose concentration, weight of residual potential sugar per weight of slurry; and

(b) Decomposition of glucose (B $\longrightarrow$ C)

$$r_2 = 4.9 \times 10^{14} C^{0.55} \ \exp(-32,800/RT)C_B ,$$

where C_B = sugar concentration, weight of sugar per weight of slurry. The sugar depends upon time, temperature, and acid concentration.

Chemical hydrolysis can be done either on a batch or on a continuous basis. The main difference between the two approaches is in the equipment. For the batch process only a single reactor (digester) is used. All of the steps in the hydrolysis take place in the single unit. In a continuous system, more than one reactor is used. In general, the continuous process resembles the batch process and incorporates the same processing steps, namely, hydrolysis, flash vaporization, neutralization, and centrifugation. The main difference is in the reactor. The continuous reactor system consists of a series of individual reactor tubes with a screw press after each unit. The first reactor is designed to hydrolyze only the hemicellulose portion of the charge. The sugars formed in this step are removed by passing the discharge from the reactor through a screw press. The sugars are in the liquid portion. The pulp (solids) fraction is reacidified and passed through a second and a third reactor. Some systems include a fourth reactor. The reactions after the first one are designed to hydrolyze the alphacellulose portion of the charge.

The batch system is more economical for small operations. The top capacity of a small system would be on the order of a total of 125 tons (113 metric tons/day) based on two 80-ton/days using the same storage tanks and possibly the centrifuge, with a resulting capacity-cost exponent close to 1.0.[157] The principal disadvantage of the batch system is the absence of economies of scale-up beyond the 125 tons/day.

The yield of sugars ranges from about 35 to 45 percent of the incoming cellulose (i.e., 5,600 to 7,200 lbs. [2,540 to 3,266 kg] from an 80-ton [72 metric-ton]/day plant). The estimated manufacturing cost in 1968–1969 ranged from $0.0555/lb. ($0.122/kg) with an 80-ton/day plant to $0.0365/lb. ($0.0805/kg) with a 1,000-ton/day plant.[157] (At the time of this writing [January 1976], the numbers would be $0.081 to $0.053/lb. ($0.179 to $0.116/kg). The fraction of these total costs attributable to providing the waste feed ranges from 30 percent at 80 tons per day to 44.5 percent at 1,000 tons per day (907 metric tons/day). These cost estimates most likely are optimistic in that those for size reduction are underestimated. Nevertheless, at today's prices for sugar the process begins to seem potentially attractive, even with the allowances made for a drastic increase in size-reduction costs.

Biological (Enzymatic) Hydrolysis

A synonym for "biological hydrolysis" is "enzymatic hydrolysis." The latter term is more commonly used because the hydrolysis of the cellulose is through enzymatic reactions. Nevertheless, the term "biological" also is valid because the enzymes are synthesized by living organisms, bacteria and fungi (yeasts).

■ **Chemistry of Enzymatic Cellulose Hydrolysis:** The description of the chemistry of enzymatic hydrolysis of cellulose which follows is largely based upon one given in a report by Rosenbluth and Wilke.[158]

Since, as mentioned earlier, the crystalline regions are relatively inaccessible to the large enzyme molecules, the rate of enzymatic hydrolytic activity is highly dependent upon the crystallinity of the substrate. Other factors that influence the biological hydrolysis of cellulose are the surface-to-volume ratio and the presence of associated substances occurring with cellulose in nature. The most important among the latter is lignin. The lignin, which is present in large amounts in wood, protects the cellulose from enzymatic attack by lessening the accessibility of the cellulose.

- *Mechanism* – A wide variation exists in the hydrolytic activity of the many groups of microorganisms capable of degrading cellulose. Each group produces an enzyme system that possesses characteristics unique unto that group. Some microorganisms produce cellulases capable of hydrolyzing only soluble derivatives of cellulose, while others synthesize cellulases that are very effective in hydrolyzing native cotton. The existence of these two types of capabilities has led to a method for dividing cellulase enzymes into two separate groups, namely, C_1 and C_x. The C_x enzyme acts to produce linear anhydroglucose from cotton cellulose (about 70 percent crystalline) and other highly crystalline forms of cellulose. The C_x enzyme hydrolyzes these linear chains to form soluble carbohydrates, usually cellobiose and glucose. It also acts to hydrolyze reprecipitated or substituted forms of cellulose, such as carboxymethyl cellulose. The action of the two enzymes is illustrated in Figure 23. Others refer to the C_1 enzyme as the "affinity factor." The other member enzymes of the cellulase complex are the hydrolases and a beta-glucosidase.

A wide variation exists between cellulolytic microorganisms with respect to the total and relative amounts of C_1 and C_x activities in their enzyme systems. Since the initial step, involving C_1, is the slowest one, it determines the rate of hydrolysis of resistant cellulose. The microorganism thus far found to be most effective in hydrolyzing cellulose is *Trichoderma viridae*. Its effectiveness is the result of the production of an enzyme system characterized by a high degree of C_1 activity. It is the only group that has a cellulase system highly active on crystalline cellulose. Consequently, in designing a hydrolysis plant, it generally is the selected fungus.

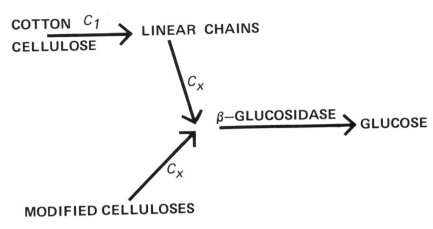

Figure 23: The mode of action of cellulolytic enzyme systems

The enzymatic hydrolysis generally occurs outside the microbial cell. This follows from the fact that most cellulases are extracellular in nature. However, certain bacteria, the cytophaga, are believed to have the enzyme system tied up in the cell wall; and, consequently, contact must be had between the cellulose and the cell wall for hydrolysis to take place. The cellulase complex is constitutive in some microorganisms and is induced in others. Cellobiose is the principal inducing agent, although if present in a sufficiently heavy concentration (i.e., 0.5 to 1.9 percent), it can act to repress production and inhibit activity. Generally, some cellobiose is present with cellulose, and consequently fungal growth is assured. The simple addition of cellobiose to a culture to accelerate enzyme production would serve no purpose, inasmuch as the additional material would be utilized as a nutrient by the microorganisms. However, if the growth of the microorganisms is simultaneously inhibited, then the yield of the enzyme would be enhanced. High concentrations of glucose also result in the inhibition and repression of cellulase formation. The role of products of enzymatic action on the induction of cellulase is diagrammed in Figure 24.

■ **Process Description:** Inasmuch as until and including the present time enzymatic hydrolysis has been and is carried on at most on a small pilot-plant scale, process designs are largely

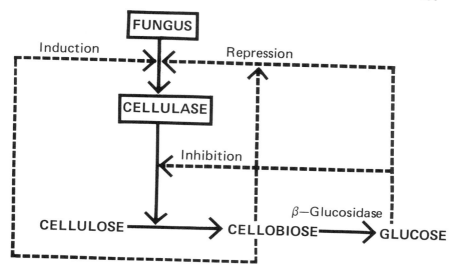

Figure 24: The role of products in the induction of cellulase in *Trichoderma viridae*

tentative in nature. However, when a full-scale operation eventually comes into being, it will of necessity include certain basic design factors. The general process description given by Rosenbluth and Wilke[158] takes those factors into account. It is diagrammed in Figure 25.

As the figure shows, the process is divided into several steps and includes two basic inputs, namely, nutrients for the fungus and the cellulosic wastes to be hydrolyzed. The nutrients supply the nitrogen and other nutritional requirements not satisfied by the cellulosic wastes. Consequently, a first step involves nutrient ("feed") preparation. The nutrients are dissolved in water and the nutrient medium is then heat sterilized to prevent the development of unwanted microorganisms in the medium. While these two steps are in progress, the cellulose wastes are heated to 200°C and then size-reduced to a particle size of about 50 microns. Fine milling is required to break up the crystalline structure of the cellulose. The need for so doing is shown by the fact that sugar recovery from many cellulose substrates is only 6 to 7 percent at large particle sizes and 70 percent with finely milled cellulose. Studies by Mandels *et al.*[159,160] indicate that the greatest recovery

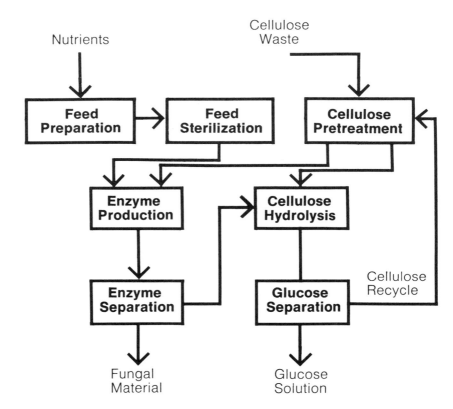

Figure 25: A schematic diagram of a process to enzymatically saccharify cellulosic waste

of sugars is obtained when the cellulosic wastes have been ball-milled. Ball-milling results in an extensive size reduction, greatest bulk density, and maximum susceptibility. Hammer-milling, fluid energy milling, colloid milling, and alkali treatment proved to be much less satisfactory.

The enzyme production step, which is the major one in the process, involves microbial growth and subsequent enzyme production. The fungus is cultured as a submerged culture in a fermenter unit equipped for mixing and aerating the culture. Here occur simultaneously the attainment of high levels of cellulase and the establishment of a cellulose-degrading enzyme system which is induced in the presence of cellulose. Rosenbluth and Wilke estimated that it would take four fermenter units of 52,000 gallons ($197m^3$) each to process 10 tons (9 metric tons) of pretreated cellulose per day. Aeration is accomplished by injection and is kept at the maximum needed to maintain an adequate oxygen transfer. The separation of the enzyme solution from the fungal mycelial mass can be accomplished by means of continuous filtration or by centrifugation. The filter cake, consisting mainly of cell material, is sent to a dryer, and the enzyme-rich filtrate is fed continuously into the hydrolysis vessel.

The cellulose hydrolysis or saccharification step is the most important one in the process. In it, the enzyme solution produced in the previous step comes into contact with the pretreated cellulosic waste. The enzyme solution catalyzes a hydrolysis of the solid cellulose to a glucose product. The product stream (about 5 percent glucose) is continuously withdrawn from the unit. Further growth of the fungus in this step is prevented by elevating the temperature of the mixture to 50°C.

Finally, the glucose solution is separated from the unhydrolyzed cellulose by means of filtration. The glucose solution may be used as a carbon source in the production of yeast as a feedstuff for animals or humans or for the production of ethanol through fermentation. The residue is dried, reground, and returned to the hydrolysis vessel. The recycled cellulosic material must be ground repeatedly to render the residual cellulose more susceptible to hydrolysis. Cell material filtered from the culture broth (enzyme solution) is passed through a dryer.

Economics of Hydrolysis

Although estimates of the hydrolysis production costs of sugar are abundant enough in the literature, they are marred by two defects: (1) They are based only on laboratory and small-scale laboratory operations, and (2) they are made by proponents of the process and consequently suffer from a certain degree of lack of objectivity, albeit unintentional and unaware. A serious cost factor among several is, as stated before, the need to finely mill the raw material. Accomplishing the extremely fine size reduction involves a technology presently not geared to processing cellulosic materials, as well as sizeable equipment and energy expenditures.

The ball-milling proposed for the process is based on designs originally developed primarily to process brittle material and not pliable substances such as cellulosic wastes.

FERMENTATION

Single-Cell Protein Production

The organism most commonly used in single-cell production on the hydrolysis products from cellulose is the yeast *Candida utilis,* also known as *Torula utilis.* Torula is selected because it exhibits rapid growth rates, has a broad tolerance in terms of environmental conditions, and can assimilate a wide variety of carbon sources, especially the pentoses. It also has been demonstrated to be a good food and fodder yeast. Bacterial cells are not selected because of their high nucleic acid content and their small size, which makes them more difficult than yeasts to harvest. In addition, certain bacteria have endotoxins and can promote allergenic reactions in humans who consume them.

The material in this section is based largely on information obtained from references 157 and 158.

■ **Process Considerations:** The fermentation process consists of the propagation and recovery of a microorganism, namely, *Candida* (or *Torula*) *utilis.* This is accomplished by growing the yeast on the sugars resulting from hydrolysis. In addition to the hydrolysate sugars, which serve as a carbon source for the yeast, other nutrients must be supplied. The major nutrients are nitrogen (as an ammonium compound), phosphorus (as a phosphate), and potassium (as K_2SO_4 or KOH). Sulfur requirements are met through the addition of K_2SO_4 or $CaSO_4$. Trace amounts of magnesium and iron are needed if they are not already present as impurities in the feedstock water. The amounts of the macronutrients are determined on the basis of the composition of the average yeast cell, namely, C, 44.6 percent; N, 8.5 percent; P, 1.1 percent; K, 2.2 percent; and S, 0.6 percent. The expected yield of dry cell mass is from 45 to 55 percent of the sugar consumed. For design purposes, a production equal to 50 percent of the sugar consumed may be assumed. Growth is carried out within the temperature range of 20 to 35°C in a reactor equipped to agitate the culture.

Since the desired fermentation is an aerobic one, oxygen must be supplied. Typically, 1.05 lbs. oxygen are needed per pound (1.02 kg O_2/1 kg) of dry cell mass. Of course, the amount of oxygen that must be introduced into the fermenter system is determined by the rate of oxygen transfer from air to the fermentation broth. In terms of yeast production on an industrial scale, experience indicates that a rate of 120 millimoles of oxygen absorbed per liter-hour (about 3.84 grams oxygen/liter-hour) is an economic one with respect to power consumption. The corresponding yield of yeast would be (3.84) (100/105) = 3.66 grams yeast per liter-hour.

In a continuous system, the production rate is equal to the product of the cell concentration times the dilution rate. Dilution rate is here defined as the ratio of medium feed-rate to fermenter volume. The maximum cell concentration attained with undiluted feedstock equals the sugar concentration of the hydrolysate (grams/liter) times the efficiency of utilization of sugar. For example, a 5 percent concentration of sugar would result in a 250 grams/liter concentration of cellular mass. Based upon the potential yield of yeast at an oxygen transfer of 120 milli-moles/liter-hour, the dilution rate with a 5 percent sugar hydrolysate would be (3.66) (100/25) or 0.146^{-1}.

The total culture volume during steady-state operation is mass of total potential hourly production of cells per hour divided by the mass per liter as based on rate of oxygen transfer (i.e., 3.66 grams/liter-hour). In the calculation of the amount of oxygen required, the usual experience in yeast production indicates an oxygen absorption efficiency of only about 15 percent. Therefore, this factor should be taken into consideration when determining the total amount of air needed for aerating the culture. When determining the required reactor volumes, one should make an allowance equal to an additional 200 percent for expansion of culture volume due to gassing and for freeboard for foaming. The equation is

$$V = \frac{(M)\ (e)}{m} \times 3,$$

where V = reactor volume in liters, M = total mass (grams) of sugar produced per hour, e = conversion efficiency of sugar to cell mass (usually 50 percent), and m = the potential yield per liter based on rate of oxygen transfer.

The yeast cells can be harvested by settling, by filtration, and by centrifugation. As a result of the large settling area required because of the low settling velocity of yeast cells (about 1.1×10^{-5} cm/sec.), harvesting by settling is impractical. Filtration also has been found to be infeasible because of the tendency of the yeast cake to clog the filter and thereby lower the filtration rate excessively. Thus, centrifugation remains the most practical approach. Generally, a battery of three centrifuges is utilized. The solids concentrate from the first centrifuge is washed to remove materials that might impart an undesirable taste or odor to the product. The solid paste from the third centrifuge can be dried by means of a double drum atmospheric dryer. Finally, the dried product is pulverized, screened, and stored.

The estimated 1968–1969 cost per pound of yeast produced ranged from \$0.06 to \$0.09/lb. (\$0.13 to \$0.20/kg) with an 80-ton/day (72 metric-ton/day) plant (basis: feed to acid hydrolysis plant) to \$0.04 to \$0.06/lb. (\$0.09 to \$0.13/kg) with a 500-ton/day plant.

Ethanol Production

The concept of converting cellulosic wastes to an energy source in the form of ethanol through the fermentation of the sugars from cellulose hydrolysis is a relatively recent one. Its origin was coincident with the onset of the realization of the shortage of fossil fuels. Because of the lateness of the interest, the existing literature on the subject was exceedingly sparse at the time this book was written.

The process involves the culture of the yeast *Saccharomyces cereviseae* under anaerobic conditions on sugars from the hydrolysis of cellulose. Wilke and Mitra[161] state that sugars resulting from the hydrolysis of cellulosic wastes have been found to be readily fermented by *S. cereviseae* at a theoretical conversion to ethanol efficiency of 83 percent. For the conversion of newsprint to ethanol, they estimate a gross energy conversion efficiency of 47 percent for the hydrolysis step and of 37 percent for the waste to ethanol step. The respective net efficiencies are 34 percent and 27 percent. The conversion efficiency is based on the heat of combustion of the feed. The gross value assumes credit for unused waste heat and present an idea of the efficiency that might be attained by adjusting the waste composition so that no unused residual solids are produced. The weakness of these projections is in the assignment of unrealistically low energy expenditure to size reduction. The problems associated with size reduction are compounded when newspaper is used as the source of cellulose.

DIRECT PRODUCTION

Most of the studies on direct production have dwelt upon the use of bagasse as the waste raw material.[155,156] The direct production of single-cell protein microorganisms is a one-step process without the intervention of saccharification. Consequently, the organisms that constitute the feedstuff product must be cellulolytic to be able to use cellulose as a carbon source. Usually cellulose biodegradation by such microorganisms is a slow process measured in terms of days. However, it has been found that the rate of degradation can be dramatically increased by utilizing one of the more recently isolated rapid-growing bacteria or fungi and by simultaneously modifying the cellulose structure, i.e., reduce its crystallinity in such a manner as to enhance microbial utilization.

An important property of cellulose in this respect is that it can be penetrated and swollen by certain strongly electrolytic solvents, one of which is NaOH. Accomplishing a swelling of the cellulose through exposure to a mild alkaline solution (2 to 8 percent NaOH) followed by an air oxidation results in a decrease in relative crystallinity of cellulose, lowers the degree of polymerization, and disrupts the physical structure of lignin that may be in the wastes. The swelling also causes the lignin sheathing to be disrupted, thus making the cellulose more accessible to enzymes. Aeration brings about the depolymerization.

Treatment of bagasse in such a manner can bring about an increase in soluble carbohydrate content from an initial 2 percent to almost 18 percent and a lowering of overall crystallinity from an original of about 50 percent down to 10 percent.

An example of a process involving direct production is one developed by researchers at Louisiana State University.[156] The organisms used by the researchers were *Cellulomonas flavigena* and *C. uda.* They report the doubling time for the *Cellulomonas* spp. to be about 3.5 to 3.7 hours. A flow diagram of the process to handle bagasse is presented in Figure 26. Nutrients added to the wastes slurry are the same as those in the enzymatic hydrolysis and in cell production. The equilibrium cell density is about 0.5 grams/liter.

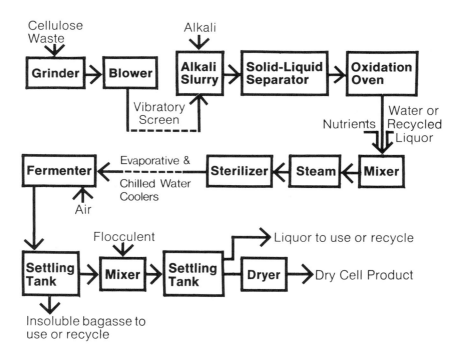

Figure 26: Direct production of single-cell protein from bagasse

The conversion of cellulosic wastes directly into single-cell protein is attended by many problems. One is that the micro-organisms used in the fermentations must possess a high degree of cellulolytic activity and must be able readily and efficiently to use the breakdown products as a source of nutrient. A second difficulty comes from the need to modify conventional submerged culture fermentation to accommodate a solid substrate. A third problem is that the amino acid composition of the protein content of the organisms to be produced must be adequate for its use as a single-cell protein. It is possible to alter the amino acid composition of the protein to some extent by inducing the organisms to produce an intracellular enzyme in large quantities. A fourth and very difficult problem is in the type of protein obtained from microbial cells. From 40 to 50 percent of the dry weight of microbial cellular matter is proteinaceous in nature. This actually could be regarded as an asset except that 20 to 25 percent of the dry weight is in the form of nucleic acids. This latter poses a problem when microbial cells are used directly as animal feed.

5. PHOTOSYNTHETIC RECLAMATION OF AGRICULTURAL WASTES THROUGH ALGAL CULTURE

INTRODUCTION

The subject of the photosynthetic reclamation of wastes through the culture of algae can be classified basically in the category of either wastewater treatment or solid waste treatment. The research conducted in the early 1950s at the University of California (Berkeley) was directed primarily to the application of the principles of photosynthetic reclamation to municipal wastewater (sewage). It was only later in the course of the research that the application of the principles to the treatment of other organic wastes began to be explored by the University researchers. However, the principal potential of the system of photosynthetic

206

recovery as developed at the University continues to be in the field of wastewater management.

Since the emphasis in this book is on solid wastes, the presentation on photosynthetic reclamation is limited herein to agricultural wastes, and more specifically to manures. Furthermore, it deals only with those systems that involve the production of a harvestable crop of algae simultaneously with water reclamation. Therefore, the conventional "oxidation" or "facultative" pond approach is not covered. Also, the coverage is brief and is intended mainly to call attention to the existence of the approach.

PRINCIPLES

The basic phenomenon upon which the method is based is the conversion of visible light energy into the chemical energy of cellular mass through the agency of chlorophyll in green plants. In this case the green plants are algae dispersed in water. It is the utilization of an algal culture that distinguishes the method as described herein from, for example, the use of manures in conventional agricultural crop production. The mechanism by which wastewater treatment and reclamation of the nutrients contained in the waters is accomplished is a loose form of symbiosis between bacteria and algae. The bacteria decompose complex organic molecules into simple compounds that can be assimilated by the algae. Another product of bacterial activity useful to algae is CO_2. The algae in turn through photosynthesis release oxygen in the water. The oxygen is utilized by the bacteria to decompose the organic wastes. The relationship between algae and bacteria is illustrated by the diagram in Figure 27.

By the very nature of the relationship between algae and the bacteria, certain basic requirements in photosynthetic reclamation come to mind: (1) A fundamental and obvious prerequisite is that the waste, the bacteria, and the algae must be in an aqueous suspension. (2) Light must be available to the algae. (3) Since the process is fundamentally biological, temperature and other environmental factors must be at a level suitable to the needs of the algae and the bacteria. (4) Toxic substances should not be present in concentrations inhibitory to the algae and the bacteria.

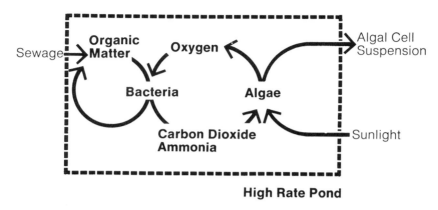

High Rate Pond

Figure 27: The algae-bacterial "symbiosis" in the photosynthetic treatment of organic wastes

The requirement for an aqueous suspension implies that the wastes (i.e., here, manures) must be slurried. The permissible solids concentration of the slurry is determined by the light requirement.

The light requirement brings with it a number of constraints, among which are: (1) Pigmentation imparted to the suspending water must not be such as to significantly filter out light impinging upon the algal suspension. (2) Similarly, the suspended solids concentration must not be so great as to screen out light to a significant degree. In other words, turbidity is to be minimized. (3) The depth of the culture must not exceed that to which light can penetrate, i.e., the culture must be shallow. (4) With special exceptions, the environmental temperature limits the practical application of the process to regions where climatic conditions are favorable. A major exception would be where access is had to thermal waste discharge, e.g., heat dissipation from cooling water from an industrial plant. Another would be where waste treatment need not be continuous throughout the year. There, the process could be operated during the warm months of the year.

■ **Microorganisms**: At least until the present, the principal algal types in nutrient recovery through algal culture have been and are unicellular green algae. The unicellular green algae were selected mainly because of their rapid growth rate, wide tolerance of environmental conditions, and ease of dispersal in the liquid medium. The principal genera are *Chlorella, Scenedesmus,* and, to a lesser extent, *Euglena, Chlamydomonas,* and *Microactinum.*

In recent years efforts have been initiated to find a filamentous type of algae that can match the unicellular green algae in good points and yet retain the principal advantage of filamentous forms, namely, ease of harvesting. The search has generally centered on the nontoxic blue greens. At the time of this writing, a great deal of work was being done on the culture of *Spirulina.* The filaments of *Spirulina,* although minute, are of such a nature as to promote the ready settling of the organisms out of suspension in a quiescent culture. Unfortunately, many problems related to their culture remain to be solved before *Spirulina* can supersede the unicellular green algae in waste treatment.

Little attention has been given to the bacterial members of the "symbiosis," probably because conditions which have been demonstrated as being optimum for the algal members apparently are also suitable for the bacteria. At any rate, this author has yet to come upon a case involving an open culture of algae in a waste medium in which impairment of bacterial activity was the limiting factor. The bacterial groups are those present in the wastes, and perhaps airborne "contaminants."

■ **Environmental Factors:** The principal environmental factors can be grouped under light, climate, and nutrition.

● *Light* – As far as light is concerned, analyses indicate that sunlight is the only economically and energetically feasible source of light energy.[162,163,164] It is the visible portion of sunlight that is used by the algae in carrying on protosynthesis. An idea of the daily quantities of visible light penetrating a water surface can be gained from the data in Table 17. Generally, light impinging upon the pond surface is in excess of the needs of the algae present at the concentration characteristic of an outdoor culture. However, the depth to which the light penetrates the culture is a function of the pigmentation and turbidity of the culture liquid. Incidentally, excessively intense light irradiation can cause damage to the chlorophyll molecule ("bleaching"), thereby inhibiting algal growth. Efficiency of light-energy conversion decreases as light intensity exceeds light saturation, i.e., about 200 to 600 foot-candles for most species. The efficiency

TABLE 17. VISIBLE SUNLIGHT ENERGY
AT THE EARTH'S SURFACE (CAL CM^{-2} DAY^{-1})

Latitude	March		June		Sept.		Dec.	
	Min.	Max.	Min.	Max.	Min.	Max.	Min.	Max.
0°	206	271	108	236	207	269	195	258
20°	168	246	148	284	176	252	120	182
40°	85	161	171	296	112	203	24	66
60°	33	107	174	294	38	126	1	5

increases with decrease in light intensity down to about 100 to 200 foot-candles. The level varies with the algal species.[165-167]

For a practical operation, maximum efficiency in light utilization is not necessarily the optimum for the process as a whole. The reason is that the maximum amount of biomass and oxygen yield is not always coincident with maximum efficiency of light-energy utilization. For example, the light intensity reaching a culture may be at the level at which the algal cells are converting light energy at maximum efficiency, and yet the total light energy penetrating culture would be so low as to be sufficient to meet the energy needs of only a very low concentration of algae. The resulting yield of biomass per unit of culture area and of culture volume processed would be so small as to render the operation economically and energetically infeasible. Usually, where the algae are raised in pond culture, culture performance in terms of biomass yield and oxygen production for waste stabilization is sufficient if at any instant two-thirds of the total mass of algal cells receive sufficient illumination.

• *Climate* — As stated before, temperature is one of the key factors that determine the geographical locations at which the method is economically feasible. Using economics as the constraint, the method is limited to those regions of the world in which the production pond would be covered with ice for only about one month out of the year. The economics is based on land and facility usage. Temperature exerts its effect on microbial activity — bacterial and algal. As the temperature approaches freezing, biological activity slows down and, for all practical purposes, ceases at a few degrees above 0°C. Limits at the higher temperatures are a function of the algal species. With the exception of a few thermophilic strains, algae begin to be severely adversely affected when the temperature begins to exceed 35°C and are killed when it rises above 40 to 42°C.

The effect of climatic conditions other than temperature is by way of the occlusion of sunlight when the sky is overcast. The effect of air movement (wind) is a certain amount of mixing in outdoor pond cultures.

● *Nutrition* – The essential macronutrients for algae are carbon and nitrogen. An exception to this generalization is constituted by the nitrogen-fixing algae. Elements that fall somewhere between the "macro" and the "micro" categories are phosphorus, iron, magnesium, and calcium. The micronutrients are the same as those for microbes in general. Unless CO_2 is injected into the pond culture by special means, the two sources of CO_2 are the metabolism (oxidation) of organic matter by the bacteria and the diffusion of CO_2 from the ambient air into the culture. Algae normally use CO_2 as the carbon source, although some types can also use the bicarbonate ion. In an outdoor pond, lack of carbon often is the limiting factor. A side effect of carbon utilization is a rise in pH level of the culture. This occurs because frequently the carbonates are the main buffer compounds in a culture.

The ammonium form of nitrogen (NH_4-N) is the one most readily utilized by algae. Not far behind are the nitrate and urea forms. Certain of the amino acids also can be used as direct sources of nitrogen. In a symbiotic situation in which wastes serve as the nitrogen source, NH_4-N usually is the form in which nitrogen is assimilated. The actual phosphorus content of algal cells is quite low. Nevertheless, the amounts added are usually much in excess of that assimilated. The larger than necessary dosages compensate for losses due to precipitation brought about by the high pH level characteristic of an unbuffered culture during periods of active photosynthesis.

The source of the macro- and micronutrients is the organic matter in the wastes. If animal wastes are the type being treated, all of the essential nutrients are present and no further additions need be made. The major exception, especially in the presence of an abundance of light, may be lack of carbon in a form available to the algae. Carbon may be supplied in a bicarbonate form, but preferably by diffusing CO_2 into the culture, although the latter is attended by many mechanical difficulties.

An optimum pH level most likely is between 6.5 and 7.5, since the range spans the optimum level for most bacteria and algae. It is also within the range at which the nutrient elements are soluble and hence available to the microorganisms. As the pH level rises above 7.0, insoluble complexes, especially of phosphorus and of iron, begin to form and precipitate out of solution. Unfortunately, pH is difficult to regulate in an outdoor pond except by applying expensive controls. The shift results from the uptake of CO_2 by the algae and a consequent shift to the right of the following equations:

$$CO_2 + H_2O \rightarrow H_2CO_3 \rightarrow H^+ + HCO_3{}^- \rightarrow H^+ + CO_3{}^=$$

On a sunny day the pH level may rise as high as 10.5 to 11.0; and during summer weather it rarely drops to as low as 8.5 at night.

DESIGN

General Considerations

Inasmuch as only agricultural wastes are concerned, this presentation on wastes is centered on the outdoor open-pond culture of algae. These limitations are placed on the discussion because, as mentioned earlier, they encompass the only presently economically feasible approach to algae production from agricultural wastes. The major design considerations are depth, mixing, and retention period. From these a number of corollary factors arise.

■ **Depth:** The major factor that determines the feasible depth of an algae production pond is extent of light penetration. Obviously, the indicated depth will vary with change in culture conditions and subsequently in algae concentration. An indication of the desired depth is given by the data in Table 18.[168] The data

TABLE 18. RELATIONSHIP BETWEEN INCIDENT LIGHT INTENSITY AND DEPTH[a,b]

Algal Concentration (mg/liter)	Depth in cm for Corresponding Incident Light Intensity			
	1,000 foot-candles	2,000 foot-candles	5,000 foot-candles	10,000 foot-candles
50	23	30	39	46
100	11.5	15	19.5	23
200	5.8	7.5	9.8	11.5
400	2.9	3.8	4.9	5.8

[a] From Bogan et al.[168]

[b] Algae concentration such that light intensity at any depth is 100 foot-candles.

are based on the absorption of light energy as determined according to the Bier-Lambert law, an extinction coefficient of 2×10^{-3} cm^2/mg, and a lower light limiting range of 100 foot-candles.

As a result of their experience with outdoor ponds and using sewage as a substrate, Oswald and Golueke concluded that a depth of about 4.5 in. (11 cm) would theoretically be optimum.[169,171] They also showed, however, that in practice such a more workable depth would be from 8 to 14 in. (20 to 35 cm), depending upon incident light energy and climatic conditions.

■ **Mixing:** Mixing brings about: (1) dispersion of nutrients, dissolved gases, and microbial cells such that all are in contact with each other; (2) a resuspension of settled algae and flocculated bacterial cells; and (3) an exchange of gas between the ambient atmosphere and the culture medium. The first two responses are beneficial to algal growth, whereas the third is not always so. As a pond approaches "steady-state," a bottom sludge layer consisting of bacterial floc and some settled algae cells is built up. Normally, the depth does not exceed an inch or two (2 to 5 cm), and the layer is aerobic. During mixing this layer is suspended and oxygenated, and its nutrients are dispersed throughout the culture. During the daytime, an outdoor pond usually is supersaturated with dissolved oxygen (i.e., 20 to 30 mg/liter). Mixing causes the minute bubbles of oxygen to coalesce and eventually be released to the external atmosphere. Consequently, mixing results in a loss of oxygen. However, since oxygen is present in excess, the loss need not be a disadvantage, at least not during the daytime. An advantage may be in an increase in diffusion of ambient CO_2 into the pond culture. This transfer is enhanced by the high pH level of the pond and, during the day, by a CO_2 concentration lower than that of the atmosphere.

Mixing introduces the problem of turbidity, inasmuch as through it the bottom sludge layer is resuspended. Despite the induced turbidity, a pond must be mixed at least once each day, and possibly two times on days unusually favorable to algal growth. If a pond is not mixed, an excessively deep bottom sludge layer builds up and becomes anaerobic, and the pond eventually fails. Since the adverse effect of turbidity stems from the impedance of light penetration, the obvious course is to limit the mixing to the dark hours. The best approach has been found to be the institution of a daily one- to two-hour mixing period timed such that it ends at daybreak.

Mixing can be accomplished by recirculating the pond contents at a velocity rapid enough to suspend the settled solids (i.e., "scour" the bottom). A velocity of about 1 ft./minute (30 cm/minute) should serve the purpose. Other ways of accomplishing it are by using a broom or any other stirring device.

Because an algae production pond is best operated on a continuous basis, the factor of retention time assumes importance. To arrive at a suitable retention time, an important point to keep in mind is that a direct extrapolation from laboratory culture or small pilot scale to full practical scale would lead to a completely infeasible retention time — the time would be much too short. The reason is that providing for a full-scale culture the degree of control and favorability of conditions possible with small-scale culture is especially impractical in algae production. Thus, while *Chlorella, Scenedesmus,* and other comparable types may have a potential doubling time of four to eight hours, for an outdoor pond one must design in terms of three and four days for favorable climatic conditions, and five days or more for less suitable conditions. At a detention period of three to four days and under suitable climatic conditions, a waste stabilization equal to 120 to 170 lb. BOD (biochemical oxygen demand) acre-day can be met with 1 acre of pond at a depth of 1 ft. (134 to 187 kg/ha-day with 1 ha at a 30-cm depth).

■ **Construction:** The requirement for daily mixing, combined with the necessity of avoiding turbidity plus the shallowness of the pond, results in certain construction essentials. A primary requirement is a hard-surface bottom. An earth bottom would not suffice because, with each mixing, silt or soil would be suspended, thus adding to the turbidity caused by the resuspension of the culture's settleable solids. Moreover, the velocity required to resuspend the solids would lead to a considerable erosion of an earth bottom. The type of material used to make a hard bottom is not critical. Concrete for the sump pump and discharge apron is indicated for a large-scale operation. Asphalt at parking-lot specifications is satisfactory for the remainder of the pond. Suggested, but hitherto not tried, is a hard-packed gravel bottom for the body of the pond. A second feature, namely, the installation of dividers to channel the flow, increases the efficiency of the mixing equipment.

■ **Harvesting the Algae:** Harvesting single-cell green algae is a complex process that involves three steps, two of which can be combined. The steps are initial concentration, dewatering or secondary concentration, and final drying and processing for storage or use. The third step may be omitted in operations in which the algae product is to be fed directly to the livestock — in other words, those operations in which no storage is involved. A full description of these steps has been given in publications by Oswald and Golueke.[163,172-174]

The need for initial concentration arises from considerations of economic practicality. The relatively low concentrations (100 to 400 mg/liter) of algae encountered in even the more productive algae production ponds necessitates the processing of large volumes of liquid to obtain relatively small yields of product. By interposing an initial concentration, the volume of culture liquid or material to be processed to the storage or utilization stage is reduced to a readily manipulated amount.

After a thorough exploration of the many procedures that have been developed for the concentration of highly dilute suspensions of low density solids, Golueke and Oswald came to the conclusion that centrifugation and chemical precipitation were the only two practical approaches to initial concentration. At first sight, vacuum filtration seemed to be an attractive method; but, despite the use of a wide variety of filtration procedures and filter aids, the yields remained infeasibly low. The problem arose from the fact that the algal cells clogged the filter medium and filter-aid before an appreciable amount of algae could be collected. A careful inspection of filtration procedures proposed and claimed as being successful in recent years shows that they all involve chemical or other type of precipitation (flocculation) prior to filtration. The latter also applies to the so-called "flotation" methods.

Technologically, centrifugation is an excellent method for the initial concentration. The average industrial large-scale continuous centrifuge can well serve the purpose. The problem is one of high costs, capital as well as operational, and of an extensive energy requirement. The curves in Figure 28 show energy requirement per ton of algae removed as a function of throughput rate.[173] The machine used in the study was a Dorr-Oliver-Merco B-30 continuous centrifuge. The algal concentration of the input was 200 mg/liter. These requirements could be materially lowered by increasing the algal concentration of the pond culture perhaps to 800 or 900 mg/liter. The reason is that at low concentrations ($<$ 1,000 mg/liter), power is almost entirely a function of volume of liquid processed. It should be remembered, however, that economic feasibility places a sharp upper limit on measures that can be taken to increase pond production.

Chemical precipitation may be accomplished by raising the pH of the pond suspension to 10.5 or higher through the addition of $Ca(OH)_2$ (i.e., lime) or by adjusting the pH to 6 to 7 and adding aluminum sulfate (alum). Precipitation also can be induced by the addition of cationic polymers. While chemicals other than lime or alum can be employed successfully, Golueke and Oswald have found their use to be attended by serious disadvantages. On the other hand, lime and alum are easy to use and impart little or no undesirable characteristics to the algal product. If need be, aluminum can be removed from the concentrate by lowering the pH to 3.5. With heavily buffered waters (heavy concentration of

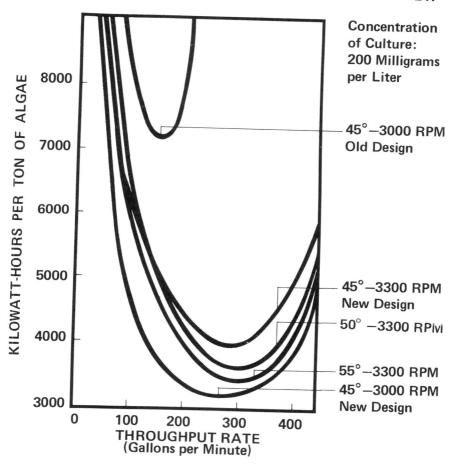

Figure 28: Energy requirement per ton of algae removed as a function of throughput rate. (Dorr-Oliver-Merco B-30 continuous centrifuge.) Algal concentration of input = 200 mg/l[173]

dissolved solids), and especially with hard waters, the required dosages of alum and lime are markedly increased. The increase for lime may be so great as to rule out its use. Under such circumstances, a cationic polymer would be suitable.

An interesting and effective method of accomplishing initial concentration involves the passing of the algal suspension through a column of strong or weak cation ion exchange resins.[174] The effluent from the strong cation column can be used to regenerate the weak cation column. Passage through the column results in a change of surface characteristics of the algal cells such that, when backwashed from the column, they agglomerate in much the same manner as occurs in the presence of cationic polymers. The effluent is sparkling clear. The only and yet decisive factor that prevents the process from becoming economically feasible is the high cost of the acid needed to regenerate the columns, namely, $200 to $300/ton dry weight of algae processed.

The product from the initial concentration step has too low a concentration (1 to 2 percent solids) for the final process step and, consequently, a dewatering step is needed. Dewatering can be accomplished with the use of a solid-bowl centrifuge, by gravity filtration, or by spreading the algae on a sand bed. If a continuous-type solid-bowl centrifuge is used, it may be necessary to precoat the machine with lime. The precoat fills the clearance between the scroll and the wall of the centrifuge, thus making the algae cake accessible to the scroll by which it is discharged from the machine. The precoat must be renewed each time the operation is interrupted and resumed because it drops from the wall when the rotor stops spinning.

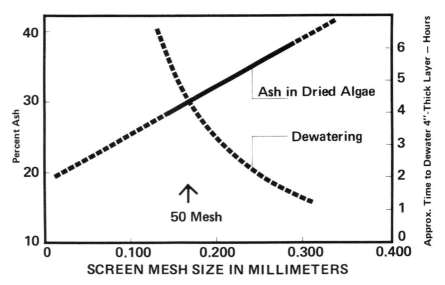

Figure 29: Ash content and dewatering time for algae dried on sand of varying mesh sizes

Vacuum filtration is not a feasible dewatering method for reasons identical to those given for filtration in initial concentration.

The use of a sand bed is the least expensive method, and it can serve as a final preparation method by allowing the algal slurry to air-dry. As the slurry dries, it takes on the form of large flakes that can be easily collected with a rake. Disadvantages with the sand bed approach are an appreciable amount of sand in the final product and a certain degree of deterioration in the quality of the product. The amount of sand and the time required to dewater an algal slurry to a 12 to 15 percent solids content are indicated by the curves in Figure 29. The sand content is indicated by the ash content over and above that normally expected of organic material. Percent ash content in excess of 10 percent is due to the presence of sand.

Final drying may be accomplished by spray-drying, by air-drying, and by drum-drying. For storage, the moisture content must be less than 12 percent. Spoilage is fairly rapid even at moisture contents only slightly higher than 12 percent. The product is deep green in color and has an odor reminiscent of that of green tea or alfalfa. It has a protein content on the order of 40 to 50 percent. Feed studies conducted with chickens, sheep, and swine as test animals indicate that with the three types of animals algae can replace soybean in terms of meeting nutritional requirements. With swine, algae can replace fishmeal, albeit less efficiently. The product is not suitable for human nutrition not only because of aesthetic reasons (grown in sewage or wastewater), but also because of its poor digestibility as well as unpalatability.

Specific Agricultural Wastes

A flow diagram of a photosynthetic system designed to recover nutrients from agricultural wastes and to provide, if it be so desired, a hydraulic waste handling system is shown in Figure 30. The design is based on one developed and tested on a pilot-plant scale by Oswald and Golueke.[163] As the diagram shows, the system includes a combination of anaerobic digestion and algae production. The major components are the animal quarters, sedimentation tank, algae growth and harvesting units, and digester. If chickens and possibly cattle are involved, the digester can be omitted — in which case, instead of being subjected to digestion, the solids could be dewatered and dried and then used as a feedstuff component for ruminants or even for chickens.

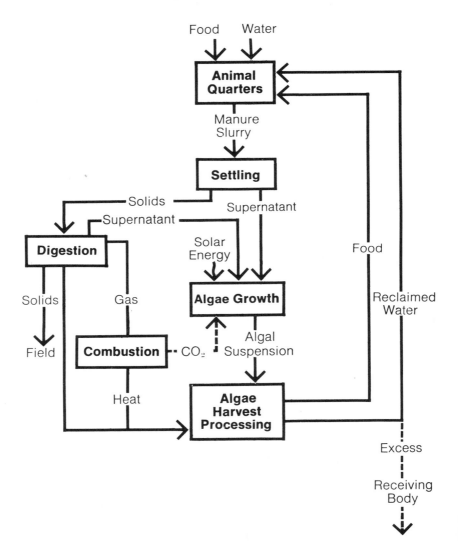

Figure 30: Photosynthetic system for agricultural wastes

The system developed by Oswald and Golueke involves a hydraulic handling of the animal wastes, during the course of which the waste solids are flushed into a sedimentation tank. The water is fed into the flushing system in an amount sufficient to keep the solids content of the resulting slurry at 3 percent or less. At such a low concentration, suspended solids are out of suspension within 15 to 20 minutes, whereas the rate of settling begins to slow down drastically as the concentration is increased above 3 percent.

As the figure shows, the supernatant is pumped directly to the algae growth pond. The solids are discharged into the digester or are processed for refeeding. In the University of California operation harvesting is accomplished by settling. Despite a too high concentration of algae in the supernatant for discharge into an external body of water, harvesting by settling would be satisfactory if all of the effluent were utilized within the system or were used for crop irrigation. At any rate, a large portion is needed for flushing and to maintain the volume of the pond at its required level. Despite the poor settling characteristics of algae, a sufficient amount settles out of suspension (20 to 30 percent) to impart a retention time to the cell mass and to meet the protein requirements of the animals in those operations in which the algae is used as a food supplement. If the effluent is to be discharged into a receiving body of water, then the algae must be removed by one of the methods described earlier.

Inasmuch as the solids from the digester are in a stabilized form, they can be disposed on the field not only without giving rise to problems, but also to the benefit of the soil.

Aside from the constraints described for algal systems in general, a major disadvantage is the surface area requirement, which stems from the requirement for light energy. Results obtained in the University of California study indicate a need for about 2 to 3 sq. ft. (0.19 to 0.28 m^2)/chicken to accomplish the necessary degree of waste stabilization, i.e., to provide the necessary oxygenation. (Cattle were not tried in the University studies.) Perhaps the surface area requirement could be reduced if oxygenation by the algae were augmented by a mechanical aeration step, especially during times of the year unfavorable for algal growth. Or, a portion of the pond could be incorporated into a facultative pond. The latter step would lessen the surface area requirement because of the greater depth (6 to 8 ft., or 1.8 to 2.4 m) permissible with facultative ponds. The system could even be made a part of a life-support system for use by family groups in developing nations situated in tropical or subtropical regions of the world.[175]

The flow diagram shown in Figure 31 is an example of such a system. The system diagrammed in the figure has never been tested on either an experimental or practical scale. However, all of the components (subsystems) have been tested in laboratory and pilot-scale studies and were found to be technologically and, quite likely, economically feasible.

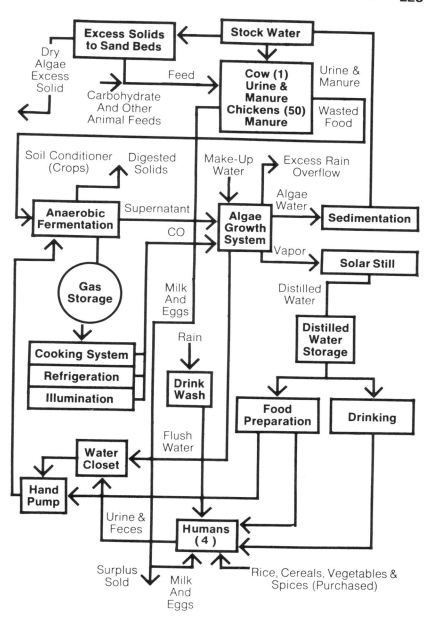

Figure 31: Schematic diagram of single-family microbiological organic waste recycle system

REFERENCES

1. Hesseltine, C. W.: "Relationships of the Actinomycetes," *Mycologia*, 52:460-474, (1960).
2. Finstein, M. S., and M. L. Morris: "Microbiology of Municipal Solid Waste Composting," *Advances in Applied Microbiology*, 19:113-151, (1975).
3. Waksman, S. A.: *The Actinomycetes: A Summary of Current Knowledge*, The Ronald Press Co., N.Y., (1967), (280 pp).
4. Kane, B. E., and J. T. Mullins: "Thermophilic Fungi and the Compost Environment in a High-Rate Municipal Composting System," *Compost Science*, 14(6):6-7, (Nov.–Dec. 1973).
5. Gray, K.: "Research on Composting in British Universities," *Compost Science*, 11(5):12-15, (Sept.-Oct. 1970).
6. Wylie, J. C.: "Waste Treatment," in *Proceedings of Second Symposium on the Treatment of Waste Waters*, Pergamon Press, London, (1961).
7. Westgate, W. A. G.: "The New Dutch Scheme for Refuse Disposal," *Public Cleansing*, 41:491, (1961).
8. Farkasdi, G.: "Experiments on the Effect of Various Additives on Windrow Composting of Ground Refuse," *International Research Group on Refuse Disposal, Information Bulletin 19*, (Dec. 1963).
9. Obrist, W.: "Experiments on the Effect of Additives on Windrow Composting of Ground Refuse," *International Research Group on Refuse Disposal, Information Bulletin 19*, (Dec. 1963).
10. Pelczar, M. J., Jr., and R. D. Reid: *Microbiology*, 2nd Ed., McGraw-Hill Book Co., N. Y., (1965), (662 pp).
11. "Second Interim Report of the Interdepartmental Committee on Utilization of Organic Wastes," *New Zealand Engineering*, 6 (11, 12), (Nov.-Dec. 1951).
12. Pfeffer, J.: "Processing Organic Solids by Anaerobic Fermentation," in *Proceedings of International Biomass Energy Conference*, Winnipeg, Canada, (May 13-15, 1973), (pp XII, 1-36).
13. Allen, M. B.: "The Thermophilic Aerobic Sporeforming Bacteria," *Bacteria Reviews*, 17:125, (1953).
14. Golueke, C. G.: "Temperature Effects on Anaerobic Digestion of Raw Sewage Sludge," *Sewage and Industrial Wastes*, 31:1225-1232, (Oct. 1958).
15. Wiley, J. S.: "Progress Report on High-Rate Composting Studies," *Engineering Bulletin, Proceedings of the 12th Industrial Waste Conference*, Series No. 94, (May 13-15, 1957), (pp 590-595).

16. Golueke, C. G., and P. H. McGauhey: "Reclamation of Municipal Refuse by Composting," *Tech Bulletin No. 9*, Sanitary Eng. Res. Lab., Univ. Calif., Berkeley, (June 1953).

17. "Composting Fruit and Vegetable Refuse: Part II," *Investigation of Composting as a Means for Disposal of Fruit Waste Solids*, Progress Report, National Canners Assoc. Research Foundation, 1133-20th St. N.W., Washington, D.C., (Aug. 1964).

18. Schulze, K. F.: "Rate of Oxygen Consumption and Respiratory Quotients During the Aerobic Decomposition of Synthetic Garbage," *Compost Science*, 1:36, (spring 1960).

19. Schulze, K. F.: "Relationship Between Moisture Content and Activity of Finished Compost," *Compost Science*, 2:32, (summer 1964).

20. Chrometzka, P.: "Determination of the Oxygen Requirements of Maturing Composts," *International Research Group on Refuse Disposal, Information Bulletin 33*, (Aug. 1968).

21. Lossin, R. D.: "Compost Studies: Part III. Measurement of the Chemical Oxygen Demand of Compost," *Compost Science*, 12:12–31, (March–April 1971).

22. Regan, R. W., and J. S. Jeris: "A Review of the Decomposition of Cellulose and Refuse," *Compost Science*, 11:17 (Jan.–Feb. 1970).

23. Trezek, G. J., and G. Savage: "Results of Comprehensive Refuse Comminution Study," *Waste Age*, 6(7):49–55, (July 1975).

24. Trezek, G. J., and G. Savage: *Size Reduction in Solid Waste Processing*, Progress Report 1973–1974, (Grant No. EPA R 801218), Mech. Eng. Lab., Dept. Mechanical Engineering, Univ. Calif., Berkeley, (1974).

25. Niese, G.: "Experiments to Determine the Degree of Decomposition of Refuse by Its Self-Heating Capability," *International Research Group on Refuse Disposal, Information Bulletin 17*, (May 1963).

26. Rolle, G., and E. Orsanic: "A New Method of Determining Decomposable and Resistant Organic Matter in Refuse and Refuse Compost," *International Research Group on Refuse Disposal, Information Bulletin 21*, (Aug. 1964).

27. Möller, F.: "Oxidation-Reduction Potential and Hygienic State of Compost from Urban Refuse," *International Research Group on Refuse Disposal, Information Bulletin 32*, (Aug. 1968).

28. Obrist, W.: "Enzymatic Activity and Degradation of Matter in Refuse Digestion: Suggested New Method for Microbiological Determination of the Degree of Digestion," *International Research Group on Refuse Disposal, Information Bulletin 24*, (Sept. 1965).

29. Lossin, R. D.: "Compost Studies," *Compost Science*, 11:16, (Nov.–Dec. 1970).

30. Allenspach, H., and W. Obrist: "Determination of the Degree of Maturity of Refuse Compost," *International Research Group on Refuse Disposal, Information Bulletin 35*, (May 1969).

31. Franconeri, P.: "How to Select a Shredder: Part I," *Solid Waste Management-Refuse Removal Journal*, 18(6):24, (June 1975).

32. Franconeri, P.: "How to Select a Shredder: Part II," *Solid Waste Management-Refuse Removal Journal*, 18(7):30, (July 1975).

33. Franconeri, P.: "How to Select a Shredder: Part III," *Solid Waste Management-Refuse Removal Journal*, 19(8):32, (Aug. 1975).

34. Gotaas, H. B.: *Composting*, WHO Monograph Series 31, World Health Organization, (1956).

35. Boettcher, R. A.: *Air Classification of Solid Wastes*, EPA Report (SW-3C), USEPA, (1972).

36. *Materials Recovery System: Engineering Feasibility Study*, National Center for Resource Recovery, Inc., 1211 Washington, DC 20036, (Dec. 1972).

37. Knapp, C. E.: "Reclaiming Municipal Garbage," *Environmental Science of Technology*, 5(10):998, (Oct. 1971).

38. Dean, K. C., E. G. Valdez, and J. H. Bilbrey, Jr.: "Recovery of Aluminum from Shredded Municipal and Automotive Wastes," *Resource Recovery and Conservation*, 1(1):55–66, (May 1975).

39. Trezek, G. J., and C. G. Golueke: "Availability of Cellulosic Wastes for Chemical or Bio-Chemical Processing," paper presented at 68th Annual Meeting of American Inst. of Chemical Engineers of Los Angeles, (Nov. 16–20, 1975).

40. Garner, G. K., and staff: "Evaluation of the Terex 74-51 Composter at Los Angeles County Sanitation District," brochure issued by Terex division of General Motors, Hudson, Ohio 44236, (May 7, 1973).

41. Senn, C. L.: *Dairy Waste Management Project: Final Report*, Public Health Foundation of Los Angeles Calif. and Univ. Calif. Agricultural Extension Service, (Dec. 1971).

42. Wylie, J. S.: "Progress Report on High-Rate Composting Studies," *Engineering Bulletin, Proceedings of the 12th Industrial Waste Conference*, Series No. 94, (May 13–15, 1957), (pp. 590–595).

43. USDA: "Composting Sewage Sludge," brochure issued by USDA: Agricultural Research Service and Maryland Environmental Service, (Sept. 1973).

44. Castro, F. J. Mena (Personnel Administrator, Direción Generál de Extensión Agricola, Toluca, Mexico): private communications, (Aug. 8, 1975).

45. Hortenstine, C. C.: *Effects of Garbage Compost on Soil Processes*, summary progress report submitted to the Bureau of Solid Waste Management, U.S. Dept. HEW, PHS, (Nov. 30, 1970).

46. Waksman, S. A.: *Humus* (ed. 2), Williams & Wilkins Co., Baltimore, MD, (1938).

47. Quastel, J. H., and P. G. Schofield: "Biochemistry of Nitrification in Soil," *Bacteriological Reviews*, 15(1) (no page numbers of reprint), (March 1951).

48. Hortenstine, C. C., and D. F. Rothwell: *Composted Municipal Refuse as a Soil Amendment*, report to the USEPA, Solid Waste Management, (1971).

49. Scarsbrook, C. E., R. Dickens, A. E. Hiltsbold, H. Orr, K. Sanderson, and D. G. Sturkie: *Conservation of Resources in Municipal Waste: Final Report*, Agr. Experimental Sta., Auburn University, Auburn, Alabama, (1970).

50. Quastel, J. H.: "Influence of Organic Matter on Aeration and Structure of Soil," *Soil Science*, 73(6):419–425, (1952).

51. Klerk, E.: "The Use of Refuse Compost in Viniculture," *International Research Group on Refuse Disposal, Information Bulletin 2*, (April 1957).

52. Nakamura, N.: "Plant Growing Experiments in Japan," *International Research Group on Refuse Disposal, Information Bulletin 17*, (May 1963).

53. Tietjen, C.: "The Utilization of Composted Domestic Refuse," in "Waste Disposal: Proceedings of the 4th International Congress of the International Research Group on Refuse Disposal," *Schweizerische Zeitschrift für Hydrologie*, 3(2):543–551, (Dec. 31, 1969).

54. Golueke, C. G.: *Comprehensive Studies on Solid Wastes Management: Abstracts and Excerpts from the Literature, Vol. I*, SERL Rep. No. 68–3, Sanitary Eng. Res. Lab., Univ. Calif., Berkeley, (June 1968).

55. Golueke, C. G.: *Comprehensive Studies of Solid Wastes Management: Abstracts and Excerpts from the Literature, Vol. II*, SERL Rep. No. 69–7, Sanitary Eng. Res. Lab., Univ. Calif., Berkeley, (July 1969).

56. Golueke, C. G.: *Comprehensive Studies of Solid Wastes Management: Abstracts and Excerpts from the Literature, Vol. III*, SERL Rep. No. 70–6, Sanitary Eng. Res. Lab., Univ. Calif., Berkeley, (Aug. 1970).

57. Golueke, C. G.: *Abstracts, Excerpts, and Reviews of the Solid Waste Literature, Vol. IV*, SERL Rep. No. 71–2, Sanitary Eng. Res. Lab., Univ. Calif., Berkeley, (1971).

58. Golueke, C. G.: *Abstracts, Excerpts, and Reviews of the Solid Waste Literature, Vol. V*, SERL Rep. No. 72–4, Sanitary Eng. Res. Lab., Univ. Calif., Berkeley, (May 1972).

59. Tietjen, C., and S. A. Hart: "Compost for Agricultural Land?" *Journal Sanit. Eng. Div.*, Proceeding Paper 650 C, ASCE, 95(SA2):269, (April 1969).

60. Orr, H. P., K. C. Sanderson, R. Self, and W. C. Marten, IV: "Utilization of Woody Plants in Containers," *Research Results for Nurserymen*, Horticulture Series No. 10, Agr. Experimental Sta., Auburn University, Auburn, Alabama, (Aug. 1968).

61. Hasler, A., and R. Zuber: "Effect of Boron in Refuse Compost," *International Research Group on Refuse Disposal, Information Bulletin 27*, (Aug. 1966).

62. *Composting of Municipal Solid Wastes in the United States*, Publication (SW–47s), USEPA, (1971).

63. Rosebury, T.: *Microorganisms Indigenous to Man*, McGraw-Hill Book Co., N.Y., N.Y., (1963).

64. Klein, S. A., C. G. Golueke, P. H. McGaukey, and W. J. Kaufman: *Environmental Evaluation of Disposable Diapers*, SERL Report No. 72–1, Sanitary Eng. Res. Lab., Univ. Calif., Berkeley, (Jan. 1972).

65. Cooper, R. C., S. A. Klein, C. J. Leong, J. L. Potter, and C. G. Golueke: *Effect of Disposable Diapers on the Composition of Leachate from a Landfill: Final Report*, SERL Rep. No. 74–3, Sanitary Eng. Res. Lab., Univ. Calif., Berkeley, (Feb. 1974).

66. Blair, M. R.: "The Public Health Importance of Compost Production in the Cape," *Public Health* (Johannesburg, South Africa), 15:70, (1952).

67. Strauch, D.: "Further Hygienic Investigations of the Influence of 'Promoting or Stimulating Factors' on Decontamination of Refuse and Sewage Sludge Components," *International Research Group on Refuse Disposal, Information Bulletin 25*, (Feb. 1965).

68. Banse, H. J., G. Farkasdi, K. H. Knoll, and D. Strauch: "Composting of Urban Refuse," *International Research Group on Refuse Disposal, Information Bulletin 38*, (May 1968).

69. Knoll, K. H.: "Composting from Hygienic Viewpoint," *International Research Group on Refuse Disposal, Information Bulletin 1*, (July 1959).

70. Joint USPHS-TVA Compost Project: *Progress Report*, Johnson City, Tenn., (June–Sept. 1967).

71. Morgan, M. T., and F. W. Macdonald: "Tests Show That *MB Tuberculosis* Doesn't Survive Composting," *Journ. Environ. Health*, 32(1):101–108, (July–Aug. 1969).

72. Van Vuren, J. P. J.: *Soil Fertility and Sewage*, Faber Press, London, (1949).

73. Golueke, C. G., and H. B. Gotaas: "Public Health Aspects of Waste Disposal by Composting," *Amer. Journ. Pub. Health*, 44(3):339, (March 1952).

74. Chinese Academy of Medical Sciences: "Sanitary Effects of Urban Garbage and Night Soil Composting," *Chinese Med. Journ.*, 1(6):407–412, (Nov. 1975).

75. Shell, G. L., and J. L. Boyd: *Composting Dewatered Sewage Sludge*, Report (SW-12c), Bureau of Solid Waste Management, U. S. Dept. HEW, PHS, (1969).

76. Peterson, M. L.: *Parasitological Examination of Compost*, Solid Waste Research Open File Report, USEPA, (1971).

77. Martin, P.: "Plant Pathology Problems in Refuse Composting," *International Research Group on Refuse Disposal, Information Bulletin 19*, (Dec. 1963).

78. Knoll, K. H.: "Public Health and Refuse Disposal," *Compost Science*, 2(1):35–41, (spring 1961).

79. Finstein, M. S., and D. V. Arent: "Composting Leaves by New Jersey Municipalities? Survey and Assessment," *Compost Science*, 15(5):6–11, (Nov.–Dec. 1974).

80. Bombard, C. J.: "Composting Leaves Saves Money in Bangor, Maine," *Compost Science*, 16(1):23–24, (Jan.–Feb. 1975).

81. Van Vorst, J. R.: "Four Leaf-Composting Communities," *Compost Science*, 13(3):18–23, (May–June 1972).

82. Van Vorst, J. R.: "Utilization of Municipal Leaves," *Compost Science*, 14(4):18–21, (July–Aug. 1972).

83. "Tenafly, N. J. Welcomes City Leaves," *Compost Science*, 5(3):16–18, (autumn–winter 1965).

84. Youdovin, S. W.: "A Two-Way Deal with Leaves," *Compost Science*, 15(5):20–23, (Nov.–Dec. 1974).

85. "The Clivus Multrum Composting Toilet," *Compost Science*, 15(5):20–23, (Nov.–Dec. 1974).

86. Gainesville Municipal Waste Conversion Authority, Inc.: *Gainesville Compost Plant*, interim report prepared for the U. S. Dept. HEW, PHS, Cincinnati, Ohio, (1969).

87. Scholz, H. G.: "Systems for the Dehydration of Livestock Wastes. A Technical and Environmental Review," in *Livestock Waste Management and Pollution Abatement*, Proceedings of the International Symposium on Livestock Wastes, Columbus, Ohio, (April 19–22, 1971).

88. Senn, C. L.: "Role of Composting in Waste Utilization," *Compost Science*, 15(4):24–28, (Sept.–Oct. 1974).

89. Babbitt, H. E.: *Sewerage and Sewage Treatment* (ed. 7), John Wiley & Sons, Inc., N.Y., (1953).

90. Fair, G. M., J. C. Geyer, and D. A. Okun: *Elements of Water Supply and Wastewater Disposal* (ed. 2), John Wiley & Sons, Inc., N. Y., (1971).

91. Vesilund, P. A.: *Treatment and Disposal of Wastewater Sludges*, Ann Arbor Science Publishers, Inc., Ann Arbor, Mich., (1974).

92. Peterson, J. R., C. Luo Hing, and D. R. Zenz: "Chemical and Biological Quality of Municipal Sludge," in Sopper, W. E., and L. T. Kardos (eds.), *Recycling Treated Municipal Wastewater and Sludge through Forest and Cropland*, Pennsylvania State University Press, University Park and London, (1973).

93. Burd, R. S.: *A Study of Sludge Handling and Disposal*, FWP1A (EPA) Pub. WP–20–4, Washington, D. C., (1968).

94. Shipp, R. F., and D. E. Baker: "Pennsylvania's Sewage Sludge Research Program," *Compost Science*, 16(2):6–8, (March–April 1975).

95. Kirkham, M. B.: "Disposal of Sludge on Land: Effect on Soils, Plants, and Ground Water," *Compost Science*, 15(2):6–10, (March–April 1974).

96. Anderson, M. S.: *Sewage Sludge for Soil Improvement*, Circular No. 972, USDA, (Nov. 1955).

97. Stone, R.: *Disposal of Sewage Sludge into Sanitary Landfill*, Final Report (SW–71d), USEPA, (1974).

98. Kudrna, F.: "Chicago's Prairie Plan — Why Does It Work and What Does It Mean to Other Cities?" *Compost Science*, 15(3):22, 23, (summer 1974).

99. Alter, J. H.: " 'Nu-Earth,' Chicago's Merchandizing Program," *Compost Science*, 16(3):22, 24, (May–June 1975).

100. Lynam, B. T.: "Methods of Liquid Fertilizer Application," in *Proceedings of Conference on Land Disposal of Municipal Effluents and Sludges*, Rutgers, USEPA publication (EPA – 902/9–73–00) (1973).

101. Reed, C. H.: "Equipment for Incorporating Sewage Sludge into the Soil," in *Proceedings of Conference on Land Disposal of Municipal Effluents and Sludges*, Rutgers, USEPA publication (EPA – 902/9–73–00) (1973).

102. Ewing, B. B., and R. J. Dick: "Disposal of Sludge on Land," in Gloyna, E. F., and W. W. Eckenfelder (eds.), *Water Quality Improvement by Physical and Chemical Processes*, Univ. Texas Press, Austin, Texas, (1970).

103. Dotson, G. K.: "Some Constraints of Spreading Sewage Sludge on Cropland," in *Proceedings of Conference on Land Disposal of Municipal Effluents and Sludges*, Rutgers, USEPA publication (EPA – 902/9–73–001) (1973).

104. Dean, R. B.: "Disposal and Reuse of Sludge and Sewage: What Are the Options?" in *Proceedings of Conference on Land Disposal of Municipal Effluents and Sludges*, Rutgers, USEPA publication (EPA – 902–73–001) (1973).

105. Hinesly, T. D., O. C. Braids, and J. E. Molina: *Agricultural Benefits and Environmental Changes Resulting from the Use of Digested Sewage Sludges on Field Crops*, Interim Report (SW–30d), USEPA, (1971).

106. Page, A. L.: *Fate and Effects of Trace Elements in Sewage Sludge When Applied to Agricultural Lands*, Report EPA – 670/2–74–005, Office of Research & Development USEPA, Cincinnati, Ohio, (Jan. 1974).

107. LaRiche, M. M.: "Metal Contamination of Soil in the Woburn-Market — Garden Experiment Resulting from the Application of Sewage Sludge," *Journ. Agric. Sci.* (Cambridge), 71:205–208, (1968).

108. Kirkham, W. B.: "Trace Elements in Corn Grown on Long-Term Sludge Disposal Site," *Environ. Sci. & Technology*, 9(8):765–768, (Aug. 1975).

109. Keeney, D. R., K. W. Lee, and L. M. Walsh: *Guidelines for the Application of Wastewater Sludge to Agricultural Land in Wisconsin*, Tech. Bull. No. 88, Dept. Natural Resources, Madison, Wis. (1975), (36 pp).

110. Keeney, D. R., and L. M. Walsh: "Heavy Metal Availability in Sewage-Sludge Amended Soils," unpublished paper, Dept. Soil Science, Univ. Wisconsin, Madison, Wis. 53705, (1975).

111. Giordano, P. M., J. J. Mordvedt, and D. A. Mays: "Effect of Municipal Wastes on Crop Yields and Uptake of Heavy Metals," *Journ. Environ. Quality*, 4(3):394–399, (July–Sept. 1975).

112. Mordvedt, J. J., and P. M. Giordano: "Response of Corn to Zinc and Chromium in Municipal Wastes Applied to Soil," *Journ. Environ. Quality*, 4(2):170–174, (April–June 1975).

113. Hinesly, T. D., R. L. Jones, and E. L. Ziegler: "Effects on Corn by Applications of Heated Anaerobically Digested Sludge," *Compost Science*, 13(4):26–31, (July–Aug. 1972).

114. Jones, R. L., T. D. Hinesly, and E. L. Ziegler: "Cadmium Content of Soybeans Grown on Sewage-Sludge Amended Soil," *Journ. Environ. Quality*, 2:351–353, (1973).

115. Hales, D.: "*Salmonellae* in Dried Sewage Sludge," *Environ. Health Journ.* (Journ. Assoc. Health Insp. Gr. Br.), 82(11):213–215, (Nov. 1974).

116. Pound, C. E., and R. W. Crites: "Nationwide Experiences in Land Treatment," in *Proceedings of Conference on Land Disposal of Municipal Effluents and Sludges*, Rutgers, USEPA publication (EPA – 902/9–73–001) (1973), pp 227–244).

117. Gerba, C. P., C. Wallis, and J. L. Melnick: "Fate of Wastewater Bacteria and Viruses in Soil," *Journ. Irrigation and Drainage Div.*, ASCE 101 (IR3), (Sept. 1975), (pp 137–174).

118. Wolf, R.: "Sludge in the 'Mile-High' City," *Compost Science*, 16(1):20, 21, (Jan.-Feb. 1975).

119. Lance, J. C.: "Fate of Nitrogen in Sewage Effluent Applied to Soil," *Journ. Irrigation and Drainage Div.*, ASCE 101 (IR3), (Sept. 1975), (pp 131–144).

120. Kellogg, H. C.: "Marketing Sewage Sludge," *Compost Science*, 14(4):16–18, (July–Aug. 1973).

121. Böhme, L.: "Composting of Undigested Sewage Sludge," *International Research Group on Refuse Disposal, Information Bulletin 27*, (Aug. 1966).

122. Epstein, E., and G. B. Willson: *Composting Raw Sludge*, contribution from the Agricultural Environmental Quality Institute, Agric. Research Service, USDA, Beltsville, Md. 20705, (copy lacks date – probably 1975 would be correct).

123. Gaby, W. J.: *Evaluation of Health Hazards Associated with Solid Waste/Sewage Sludge Mixtures*, EPA Report, EPA–670/2–75–023, National Environmental Research Center, USEPA, Cincinnati, Ohio 45268, (April 1975).

124. Hathaway, S. W., and R. A. Alexsy: "Improving Fuel Value of Sewage Sludge," *News of Environmental Research in Cincinnati*, USEPA, (Oct. 31, 1975).

125. Singh, R. B.: *Bio-Gas Plant*, Gobar Gas Research Sta. Ajitmal, Etawah, (U.P.), India, (1971).

126. Golueke, C. G., and P. H. McGaukey: "Waste Materials," *Annual Review of Energy*, Vol. 1, Annual Reviews, Inc., Palo Alto, Calif., (1976), (pp 29–49).

127. Babbitt, H. E., B. J. Leland, and F. H. Whiteley, Jr.: "The Biological Digestion of Garbage with Sewage Sludge," *University of Illinois Bulletin*, Vol. 24, No. 24 (Bulletin No. 287), Univ. Ill. Engineering Experiment Station, Urbana, Ill., (Nov. 1936).

128. Taylor, H.: "Garbage Grinding at Goshen," *Engineering News Record*, 127:441, (1941).

129. Ross, W. E., and S. F. Tolnan: "Garbage Grinding Pays its Way," *Public Works*, 84:70, (May 1953).

130. Ross, W. E.: "Dual Disposal of Garbage and Sewage at Richmond, Indiana," *Sewage and Industrial Wastes* (presently *Journ. Water Pollution Control Fed.*), 26:140, (Feb. 1954).

131. Klein, S. A.: "Anaerobic Digestion of Municipal Refuse," in *Comprehensive Studies of Solid Wastes Management: Third Annual Report*, SERL Rep. No. 70-2, Sanitary Eng. Res. Lab., Univ. Calif., Berkeley, (June 1970).

132. Klein, S. A.: "Anaerobic Digestion of Municipal Refuse," in *Comprehensive Studies of Solid Wastes Management: Final Report*, SERL Rep. No. 72-3, Sanitary Eng. Res. Lab., Univ. Calif., Berkeley, (May 1972).

133. Metcalf and Eddy, Inc.: *Wastewater Engineering*, McGraw-Hill Book Co., N.Y., (1972).

134. Chan, D. B., and E. A. Pearson: *Comprehensive Studies of Solid Wastes Management: Hydrolysis Rate of Cellulose in Anaerobic Fermentation*, SERL Rep. No. 70-3, Sanitary Eng. Res. Lab., Univ. Calif., Berkeley, (Oct. 1970).

135. Finney, C. D., and R. S. Evans, II: "Anaerobic Digestion: The Rate-Limiting Process and the Nature of Inhibition," *Science*, 190(4219):1088, 1089, (Dec. 12, 1975).

136. Pfeffer, J. T.: "Temperature Effects of Anaerobic Fermentation of Domestic Refuse," *Bioeng. Biotech.*, 16:771, (1974).

137. Gossett, J. M., and P. L. McCarty: *Heat Treatment of Refuse for Increasing Anaerobic Biodegradability*, progress report of June 1–Dec. 31, 1974, Tech. Report No. 192, Dept. Civil Engineering, Stanford Univ., (Jan. 1975).

138. Pfeffer, J. T., and J. C. Liebman: *Biological Conversion of Organic Refuse to Methane*, annual report, NSF/RANN/Se/GI-39191/75/2, Dept. Civil Eng., Univ. Illinois-Urbana Report, UILU-ENG-75-2019, (Sept. 1975).

139. Diaz, L. F.: *Energy Recovery through Biogasification of Municipal Solid Wastes and Utilization of Thermal Wastes from an Energy-Urban-Agro-Waste Complex*, doctoral dissertation, Dept. Mech. Eng., Univ. Calif., Berkeley, (1976).

140. New Alchemy Inst. West: "Methane Digesters for Fuel Gas and Fertilizer," *Newsletter No. 3*, Santa Barbara, Cal. 93101, (spring 1973).

141. New Alchemist Inst.: "Gas for Fuel and Fertilizer," in Stoner, C. H. (ed.), *Producing Your Own Power*, Rodale Press, Inc., Emmaus, Pa. 18049, (1974).

142. Singh, R. B.: "The Bio-Gas Plant: Generating Methane from Organic Wastes," *Compost Science*, 13(1):20–25, (Jan.–Feb. 1972).

143. Singh, R. B.: "Building a Bio-Gas Plant," *Compost Science*, 13(2):12–16, (March–April 1972).

144. Ecotope Group: *The Anaerobic Digestion of Dairy Cow Manure at the State Reformatory Honor Farm, Monroe, Wash.*, process feasibility study prepared by Ecotope Group (P. O. Box 5599, Seattle, Wash. 98105) and Parametrix Inc. (P. O. Box 279, Summer, Wash. 98390) for State of Washington, Dept. of Ecology, (June 15, 1975).

145. Pfeffer, J. T., and J. C. Liebman: "Energy from Refuse by Bioconversion — Fermentation and Residual Disposal Processes," paper presented at the Energy Recovery from Solid Waste Symposium, Univ. Maryland, College Park, Md., (March 13 & 14, 1975), (Pfeffer and Liebman, Univ. Ill., Urbana, Ill.).

146. Klass, D. L., and S. Ghosh: "Fuel Gas from Organic Wastes," *Chemtech*, 35:689–698, (Nov. 1973).

147. Trezek, G. J., and C. G. Golueke: "Availability of Cellulosic Wastes for Chemical or Bio-Chemical Processing," paper presented at the 68th Annual Meeting of the American Institute of Chemical Engineers, Los Angeles, Calif., (Nov. 16–20, 1975).

148. *Animal Waste Management*, Cornell Univ. Conference on Agricultural Waste Management, Syracuse, N. Y., (Jan. 13–15, 1969).

149. Amer. Soc. Agric. Eng.: *Livestock Waste Management and Pollution Abatement*, proceedings of the International Symposium on Livestock Wastes, Ohio State Univ., Columbus, Ohio, (Apr. 19–22, 1971).

150. Schmid, L. A., and R. D. Lipper: "Swine Wastes, Characterization and Anaerobic Digestion," in *Animal Waste Management*, Cornell Univ. Conference on Agricultural Waste Management, Syracuse, N. Y., (Jan. 13–15, 1969), (pp 50–58).

151. Taiganides, E. P.: "Sludge Digestion of Farm Animal Wastes," *Compost Science*, 4(2):26–30, (summer 1963).

152. Schmid, L. A.: "Feedlot Wastes to Useful Energy — Fact or Fiction?" *Journ. Environmental Engineering Division*, ASCE 101 (EE5), (Proc. Paper 11647):787–793, (Oct. 1975).

153. Dunlap, C. E.: "Physical Processing," paper presented at the Cellulose Waste Seminar, Environmental Research Center, USEPA, Cincinnati, Ohio, (1969).

154. Callihan, C. D., and C. E. Dunlap: *Construction of a Chemical-Microbial Pilot Plant for Production of Single-Cell Protein from Cellulosic Wastes*, report SW–24c, USEPA, Cincinnati, Ohio, (1971).

155. Saeman, J. F.: "Kinetics of Wood Saccharification," *Ind. & Engineering Chemistry*, 37(1):43, (Jan. 1945).

156. Fagan, R. D., H. E. Grethein, A. O. Converse, and A. Porteus: "Kinetics of Acid Hydrolysis of Cellulose Found in Paper," *Environ. Sci. & Technology*, 5(6):545, (June 1971).

157. Meller, F. H.: *Conversion of Organic Solid Wastes Into Yeast: An Economic Evaluation*, report prepared for the Bureau of Solid Waste Management by IONICS, Inc., under contract PH 86-87-204, USHEW, (1969).

158. Rosenbluth, R. F., and C. R. Wilke: *Comprehensive Studies of Solid Wastes Management: Enzymatic Hydrolysis of Cellulose*, SERL Rep. No. 70-9, Sanitary Eng. Res. Lab., Univ. Calif., Berkeley, (Dec. 1970).

159. Mandels, M. H., L. Hontz, and J. Nystrom: "Enzymatic Hydrolysis Waste Cellulose," paper presented at the 8th Cellulose Conference, SUNY, Syracuse, N. Y., (May 19-23, 1975).

160. Andren, R. K., M. H. Mandels, and J. E. Madeiros: "Production of Sugars from Waste Cellulose by Enzymatic Hydrolysis — Part I: Primary Evaluation of Substrates," paper presented at 8th Cellulose Conference, SUNY, Syracuse, N. Y., (May 19-23, 1975).

161. Wilke, C. R., and G. Mitra: "Process Developmental Studies on the Enzymatic Hydrolysis of Cellulose," paper presented at the National Science Foundation Special Seminar, "Cellulose as a Chemical and Energy Source," Univ. Calif., Berkeley, (June 26, 1974).

162. Oswald, W. J., and C. G. Golueke: "Solar Power via a Botanical Process," *Mechanical Engineering*, 85:40-41, (1964).

163. Golueke, C. G., and W. J. Oswald: "Harvesting and Processing Sewage Grown Algae," *Journ. Water Pollution Control Fed.*, 37(4):471-498, (1966).

164. Dugan, G. L., C. G. Golueke, and W. J. Oswald: *Photosynthetic Reclamation of Agricultural Wastes*, SERL Rep. No. 70-1, Sanitary Eng. Res. Lab., Univ. Calif., Berkeley, (May 1970).

165. Sorokin, C., and R. W. Krauss: "The Effects of Light Intensity on the Growth Rates of Green Algae," *Plant Physiology*, 33(2):109-113, (March 1958).

166. Golueke, C. G.: "Overall Light Energy Conversion Efficiency of a High Temperature Strain of *Chlorella pyrenoidosa*," *Physiologia Plantarum*, 15:1-9, (1962).

167. Golueke, C. G.: "The Ecology of a Biotic Community Consisting of Algae and Bacteria," *Ecology*, 41(1):65-73, (Jan. 1960).

168. Bogan, R. H., O. E. Albertson, and J. C. Pluntz: "Use of Algae in Removing Phosphorus from Sewage," *Proc. Am. Soc. Civil. Eng., J. Sanitary Eng. Div.*, SA5, 86:1-20, (1960).

169. Oswald, W. J., and C. G. Golueke: "Studies of Photosynthetic Oxygenation: Pilot Plant Experiments — Progress Report No. 8," *Sanitary Eng. Res. Lab. Report*, Issue No. 9, (Jan. 1958).

170. Oswald, W. J., and C. G. Golueke: "Large-scale Production of Algae," paper presented at the International Conference on Single-Cell Protein, M.I.T., Cambridge, Mass., (Oct. 9-11, 1967).

171. Shelef, G., W. J. Oswald, and C. G. Golueke: "The Continuous Culture of Algae Biomass on Wastes," paper presented at the 3rd International Symposium on the Continuous Culture of Microorganisms, Prague, Czechoslovakia, (June 17–23, 1968).

172. Golueke, C. G., and W. J. Oswald: *Recovery of Algae from Waste Stabilization Ponds, Vols. I, II,* (IER Series 44, Issues 7 and 8), Sanitary Eng. Res. Lab., Univ. Calif., Berkeley, (Jan. 1958).

173. Oswald, W. J., C. G. Golueke, and H. K. Gee: *Wastewater Reclamation Through the Production of Algae,* Contribution No. 22, Sanitary Eng. Res. Lab., Univ. Calif., Berkeley, (Feb. 1960).

174. Golueke, C. G., and W. J. Oswald: "Surface Properties and Ion Exchange in Algae Removal," *Journ. Water Pollution Control Fed.,* 42(8):R304, (1970).

175. Golueke, C. G., and W. J. Oswald: "An Algal Regenerative System for Single-Family Farms and Villages," *Compost Science*, 14(3):12–18, (May–June 1973).

INDEX

A

acid, concentration of,
  anaerobic digestion and,
  159–161
acid hydrolysis, 191–194
acid phase, of anaerobic digestion, 147–150
actinomycetes, action of, 10
  composting and, 8–13
  growth of, limiting factors
    in, 12, 13
  presence of, detection of,
    9, 10
aeration. See also *oxygen*.
  of cannery wastes, 77, 78
  compost toilets and, 107
  as environmental factor, in
    compost, 32–36
  in Fairfield-Hardy digester,
    81, 82

forced, 76, 77
microbial growth and, 12,
  13
turning and, 71–76
of wastes, methods for,
  71–78
windrow composting and,
  33
of windrows, 71–78
aerobic composting, advantages
  of, 3–5
  definition of, 3
  disadvantages of, 5
aerobic digesters. See *digesters*.
agricultural wastes, digestion of,
  cost of, 179–182
  photosynthetic reclamation
    of, 206–223
air. See *oxygen*.
air classification system, of
  waste separation, 66

anaerobic digestion of. See
anaerobic digestion.
animal, composting and,
102–104
carbon-nitrogen ratio of,
table of, 27
composted, quality of, 112
decomposable organic,
compost stability and,
54, 55
digestion of, cost of, 178,
179
dry processing of, 63, 64
grinding of, 42–45
human, composting and,
106–108
mechanical separation of,
systems for, 64–70. See
also *sorting.*
moisture content of, 38, 39
municipal. See *municipal
wastes.*
organic. See *organic wastes.*
particle size of, composting
and, 42–45
pathogens in, 94, 95
photosynthetic reclamation
of, 206–223
separation of, 64–70
sorting of, 40–42, 62–70
treatment of. See *waste
treatment.*
wet processing of, 64
water, in mechanical digesters,
79, 80
wet processing, of wastes, 64
windrow composting, aeration
and, 33
applicability of, 79

definition of, 6
vs mechanical, 84–86
moisture and, 36–39
technology of, 70–79
temperature curve for, 47
windrows, aeration of, 71–78
shape of, 70–78
size of, 70–78
structure of, moisture
content and, 37, 38
turning of, 71–76

Y

yeast, conversion to, from
organic wastes, 189–205
costs of, 202
in fermentation, 200